PENGUIN BOOKS

THE MOLECULE HUNT

Martin Jones is the first holder of the George Pitt-Rivers Professorship of Archaeological Science at Cambridge University. He has published widely on agricultural change in late prehistory, and on ancient DNA studies of early agriculture. He was chairman of the Ancient Biomolecules Initiative research programme.

MARTIN JONES

the molecule hunt

archaeology and the search for ancient DNA

PENGUIN BOOKS

In memory of my mother
from whom I learnt to look more closely at things

PENGUIN BOOKS

Published by the Penguin Group
Penguin Books Ltd, 80 Strand, London WC2R 0RL, England
Penguin Putnam Inc., 375 Hudson Street, New York, New York 10014, USA
Penguin Books Australia Ltd, 250 Camberwell Road, Camberwell, Victoria 3124, Australia
Penguin Books Canada Ltd, 10 Alcorn Avenue, Toronto, Ontario, Canada M4V 3B2
Penguin Books India (P) Ltd, 11, Community Centre, Panchsheel Park, New Delhi – 110 017, India
Penguin Books (NZ) Ltd, Cnr Rosedale and Airborne Roads, Albany, Auckland, New Zealand
Penguin Books (South Africa) (Pty) Ltd, 24 Sturdee Avenue, Rosebank 2196, South Africa

Penguin Books Ltd, Registered Offices: 80 Strand, London WC2R 0RL, England

www.penguin.com

First published in Allen Lane The Penguin Press 2001
Published in Penguin Books 2002
4

Printed in England by Clays Ltd, St Ives plc

contents

acknowledgements

The initial encouragement to write this book arose during lunch in a Greek restaurant in Cambridge. Not only do I warmly thank my lunching companion, Graeme Barker, for setting me off on a most stimulating project, but also for coming up with a suitable title between servings of meze. This was in the summer of 1998, shortly after the completion of the Natural Environmental Research Council's Ancient Biomolecules Initiative, which it had been my privilege to chair. The Initiative brought together around fifty researchers within seventeen projects scattered around the country, all doing pioneering work in a very novel field of applied science. Interaction with that dedicated group of people provided me with the experience and insight upon which I have constantly drawn to write this book, and my sincere thanks go to them all. I extend that thanks to everyone involved in the Initiative, the International Steering Panel, the contributors to the annual Cambridge workshops, and to the staff at NERC.

Ancient biomolecule research has been going on not only in the UK, but right across the world, from America to Japan and New Zealand. Many of those international colleagues and teams I have had the privilege to meet, others I have contacted only through email. Without exception, all have been generous and forthcoming with advice and ideas, and everywhere the enthusiasm for being part of a new scientific departure shines through. Particular thanks for dealing with specific queries and supplying information in advance of publication go to Dan Bradley (Dublin), Jane Buikstra (Albuquerque), Jacques Connan (Pau), Bill Hauswirth (Florida), Rika Kaestle (Yale), Andy Merriwether (Michigan), Svante Pääbo (Leipzig), Franco Rollo (Camarino),

Yo-Ichiro Sato (Shizuoka), Anne Stone (Albuquerque) and Mark Stoneking (Leipzig).

For help with Oliver Hunt's cover illustration, thanks go to Joan Oates and Helen McDonald. The text itself benefited from a critical reading in sections by a number of colleagues. My thanks for this invaluable assistance go to Robin Allaby, Dan Bradley, Terry Brown, Matthew Collins, Geoff Eglinton, Richard Evershed, Peter Forster, Terry Hopkinson, Chris Howe, Matt Hurles, Miranda Kadwell, Adrian Lister, Svante Pääbo, Franco Rollo, Charlie Shaw, Andrew Smith, Helen Stanley and Mark Stoneking. Thanks also to Janet Tyrrell for her scrupulous copy-editing, and to my editor Stefan McGrath and my agent Clare Alexander for their intelligent engagement, support and critical encouragement throughout the project from its inception. The text would have been a much inferior thing without them.

The least tangible contributions can also be among the more important ones. Over several years in labs, conference halls, pubs and private homes, conversations and debates with a number of colleagues have been invaluable in shaping the ideas that unfold in this book. The list of such colleagues is too long to reproduce here, but I do want to mention in this context the unstinting encouragement of Lucy Walker, and my numerous conversations with Geoff Eglinton, Terry and Keri Brown, and Robin Allaby. All these have guided and nurtured my own thoughts and ideas.

I

a different kind of past

first encounters with archaeology

A visit to a museum is quite different from actually digging up the past. In a museum, conserved objects are seen behind glass, neatly arranged and labelled in their controlled environment of scholarly explanations and humidity meters. An archaeological dig is something else. Here, the past is dirty, sticky, tactile, and quite often smelly. All this came as some surprise with my first taste of archaeology in the field, three decades ago. I had somehow imagined there would be a great deal of brushing, sifting and so forth, with one of those museum objects very occasionally emerging from the dust. In reality, what our ancestors have left beneath the ground is much more varied and challenging than that.

Rather than brushing and sifting, my introduction to field archaeology involved much wielding of picks. We pushed enormous battered wheelbarrows and hurled shovel-loads of earth as the sludge from a rather wet Somerset field found its way into our boots. Along with that moist sensation came surprise at the sheer quantity and variety of things emerging out of the ground. As our spades penetrated the turf of a slight mound in the middle of this field, our finds trays quickly filled, but not only with the kinds of objects seen behind museum glass. True, there were fragments of the elaborately decorated Glastonbury Ware pottery, its dark burnished surface adorned with the kind of swirling designs we associate with Celtic art forms across Europe. It was an unusual sensation to have been the first to hold such an artefact for over 2,000 years. But there were also large numbers of discarded animal bones, not so often seen in archaeological displays, some broken up for the soup bowl and glue pot, others elegantly fashioned into the weaving equipment of the ancient village dwellers. One day one of the diggers found another trace of their everyday lives, a broken pot spilling over with small black pellets. On closer inspection these pellets proved to be cereal grains, blackened by charring. We gathered around to peer down on the curiosity, before it was taken away to be entered into the record books of that delightful antiquarian category 'small finds', which is the archaeologist's repository for coins, beads, oddments, and anything that falls outside the most familiar categories.

As we dug down, the range of surviving materials increased. By the time the adjacent turf-line was at the level of our waists, that range had broadened dramatically. The grey clays we had been removing gave way to something very different. This blackened peat below was full of all the things that had decayed from the sediments above. Our trowels teased their way through, to reveal leaves, nuts, and vast quantities of wood. As pieces of the freshly exposed peat were pulled apart, branches of birch buried for over two millennia revealed their silvery, flaky bark. Blades of grass could be seen that still seemed to be green, as if the chlorophyll within them had remained intact. Now that they had been exposed to the air, that green coloration quickly changed to brown, just as a host of other processes of breakdown and decay set in. That exposed peat was soon peppered with insect holes, as burrowing soil animals responded with enthusiasm to the opening of the larder door. Waterlogged and sealed off from the air, the organic peats had been protected from the foraging of these creatures. We archaeologists had taken that protection away and now the limbo in which these fragments of life had been suspended would soon come to an end. In the case of the wood, this change took place before our eyes. When first exposed, the wood fragments within the peat seemed solid enough. Their surface features and patterned grain were clear to see, the main distinction from modern wood being their darkened colour. Once lifted and exposed on the grassy verge beside our trench, these pieces began to shrink as the water within them evaporated. They would twist and crack, and their surfaces would become flaky. Their solidity had been an illusion, only maintained while they remained within their enclosed and waterlogged refuge.

None of this curious decay caused great waves back in the 1960s when that Somerset dig was underway. The main business of archaeology lay beyond those flaking timber fragments on the grassy verge, a place in which another set of transformations was busily taking place. The excavated finds were being washed and marked. Toothbrushes and nailbrushes were scrubbing away in bowls of cold, murky water. The task was to remove all those things that separated the dig from the museum – the dirt, the peat, the stickiness and the smell – in order that the object we had unearthed might one day inhabit a neat museum drawer, if not a plinth within a humidity-controlled cabinet.

Since the 1960s, our perception as archaeologists of what we are digging up has changed. It has become clear that what we left behind in those peaty sediments, on the earthy surfaces we scrubbed from the pots, from the bones and organic objects, and even the partially decomposed materials we can smell, may be rich in intimate clues about past lives. We would no longer reduce what lay beneath that shallow mound in a Somerset field to an assortment of cleaned pottery fragments and other durable, easily visible objects. Bit by bit, other data were gathered and analysed as part of the archaeological routine. First came the more stable biological materials, the animal bones, and then those charred cereal grains. Then methods were refined for conserving and gleaning information out of all that waterlogged wood and the other debris within the peat. Most recently of all, it has become clear that the reason any of that organic material survives at all is because even the molecules of which it is built retain a remarkable trace of its fragile and intricate structure, for thousands, even millions, of years. Those molecules can be hunted down and analysed in their own right, and have a considerable story of their own to tell.

disciplines in flux

The great changes that have happened in archaeology, in the time that has elapsed since I first got my socks wet in a Somerset field, have not taken place in isolation. A perennial human curiosity as to what traces of the past might linger on has been an important driver of those changes, but not the only one. The things we wish to know about the past have also changed. Our queries have drawn us more and more towards organic traces, and then to the molecular information within them. Back in the days of my first dig, the question in an archaeologist's mind as he or she pondered the trays of washed and marked pottery was how those durable artefacts fitted into some grand scheme of European prehistory. The swirling patterns we had found incised on those pot fragments bore some resemblance to patterns found on metal swords and mirrors recovered from prehistoric graves in mainland Europe. Arrows could be drawn across the map to link the two, implicitly tracking a distant cultural journey, a movement from mainland Europe to the British Isles, bringing with it a package of cultural ideas and practices, encapsulated in a series of elegant swirls on the side of a pot.

By such reasoning, sites such as this contributed a few pieces to a vast and intricate three-dimensional jigsaw puzzle of Europe, two dimensions of space and one of time, all knitted together by arrows across the map, linking common attributes and design features in the durable remains recovered by excavation. These, it was assumed, traced the paths of a network of cultural journeys, migrations and invasions by which prehistory could be both narrated and explained. In this way, a line could be traced back from the swirling designs on our particular pots, in use in that Somerset lake village a little over 2,000 years ago, to similar swirling patterns on metalwork recovered from the shores of Lake Neuchâtel in Switzerland, at a site called La Tène.

Stories of this kind had two weaknesses. Just around the corner was a scientific method of dating, already tried and tested but not yet a routine tool of archaeological excavation. Radiocarbon chronology was in the process of severing and disposing of a number of those arrows. It was becoming clear that migrations and invasions could not be, and were not, the only sources of change in prehistoric society. To tally with the rigid scientific chronology now emerging, archaeologists were forced to look more seriously at the possibility of indigenous change within existing communities, rather than simply drawing arrows between poorly dated objects. This is where the second weakness came in. To examine indigenous change, we really needed to know what life was like, and we did not in fact know a great deal about the ordinary lives of those who used the elegantly decorated pots that had been at the centre of archaeological attention. Such durable artefacts had dominated the whole process of inquiry. The site report would eventually reproduce page after page of illustrations of them. The potter's workplace and the stone-worker's floor were among the few features of a reconstruction drawing that would have any detail – the people, their clothes, their dwellings and their farms dissolving into a semi-impressionistic swirl in the background.

A few years after that excavation, David Clarke at Cambridge attempted to shift focus and to make some sense of the people who actually lived and farmed on the late prehistoric villages of which our excavation had revealed a part. Those same prehistoric villages, the 'lake villages' of Glastonbury and Meare, had been excavated earlier in the century by Arthur Bulleid and George Gray, and a series of weighty tomes had been produced, cataloguing and describing their earlier excavations in great detail. Clarke combed through this data, searching for patterns in space and structure and in the distribution of different kinds of artefacts across them. In the new account he assembled, the emphasis shifted right away from sequence and pottery styles to talk of huts, workshops, granaries and stables. The village had family groups and a place in the landscape. He speculated upon the farming activities in the fields around. The settlement was coming alive, and at the same time, one of Clarke's Cambridge colleagues was beginning to look more closely at the remains of living things from that ancient landscape.

Around the time of my initiation into archaeology in a Somerset field, another team was hunting down a series of earlier features that were immersed within the expanses of peat around us. Waterlogged wooden trackways had been stumbled upon by peat-cutters since the nineteenth century at least, but now they had attracted the attention of someone who recognized that they were prehistoric in date, and who would go on to commit much of his working career to tracing them and the ancient landscapes of which they were part. In 1973, a year after Clarke's exciting paper, John Coles put together a research group of archaeologists, biologists and tree-ring specialists to unlock the treasury of bio-archaeological information contained within the peat. In the same year that the Somerset Levels Project began, a small number of 'archaeological units' were formed in Britain, to rescue archaeological information threatened by development projects. The general image of rescue archaeology at the time was of a cluster of itinerant diggers, working anxiously and rapidly in the shadow of an earthmover. A few of those units took the unusual step of putting the new bio-archaeological research at the forefront of their activities. As one of those itinerant diggers, but with a natural science degree, I became one of those bio-archaeologists in the Oxford Archaeological Unit. There weren't any models for how we should work – we were in the delightful position of making it up as we went along. One of the main tasks facing the newly formed Oxford Unit was the remains of a series of farms and hamlets, contemporary with the Somerset villages discussed above, that were disappearing as the Thames gravels were quarried away. One thing we were clear about – we wanted to do a lot more than scrub and label fragments of pottery. We wanted to float and sieve for seeds, insects and bones, in order to gather the kind of biological data that could enrich the models of prehistory that David Clarke had begun to describe.

As the 1970s progressed, many excavations brought in sieving and flotation alongside the pickaxe and trowel, in order to capture some of that data. More and more organic fragments were found within archaeological sediments. Remains of food, fuel, bodies and building materials were augmented by the debris of wild plants and invertebrates that hinted at the living environments around those living settlements. Many of these required microscopic examination, and the high-power lens brought remarkable detail into view. Even where these organic fragments had been eroded, cooked, eaten away and discoloured, still, more often than not, cellular structure within them remained. In some cases even the nuclei and other sub-cellular structures were visible. Different archaeological scientists used this detail to rebuild environments, living conditions, and methods of food production and preparation.

from ancient tissues to ancient molecules

By the 1980s 'bio-archaeology' had come of age as a routine aspect of archaeological method. The arrow-laden maps of prehistoric cultures had given way to discussions of agricultural practice, house construction and the health and nutrition of ancient rural communities. Prehistoric people and their disappeared worlds were beginning to come to life. But it did not stop there – from the late 1980s another door opened on the archaeological record and what it was able to reveal to us.

The study of ancient people was increasingly concerned with the organic, living processes these new forms of evidence revealed. It was drawing closer to studies of the biological world. But biology too had been going through great changes. To get to the heart of the living world and how it operates, biologists too had expanded the range of their observations. For many years, they had looked within whole organisms to the cells and sub-cellular structures within that formed the mechanics of life. In more recent times, they moved one stage further to the molecules that made those structures work. These included the fatty substances and carbohydrates that fuelled living processes, the proteins that built living tissue and regulated biological pathways, and the molecules that encoded the instructions for all this, the DNA at the heart of each cell. By the time archaeologists were becoming proficient at digging up fragments of ancient organisms and recognizing their tissues, biologists had already progressed deep into the heart of cellular dynamics, to decipher the molecular basis of life.

Exploring the possibilities of bio-archaeology during the 1970s and 1980s, experimenting with some fairly primitive methods of flotation and sieving, and trying to make sense of countless blackened plant fragments from prehistory, we were conscious that the biology we were then introducing into archaeology was already lodged in the past. We were attempting comparative, whole-organism studies that had a lot in common with the kind of natural history that grew in the nineteenth century and blossomed in the early twentieth. They were proving extremely valuable in bringing the archaeological past to life, but at the same time what contemporary biologists were doing suggested that we could probe much deeper. What if there were molecular traces that allowed much greater precision in identification, even when the tissue was fragmented or had disappeared completely? What if these precise identifications could take us beyond species to close relatives, to individuals, even particular genes? All this was speculation, spurred on by what could be seen through our microscopes. Whatever ancient biological material we examined, it was clear that much cellular organization had survived the ravages of time. Perhaps secreted among those cells were intact biomolecules, minute time capsules each with their own record of a distant past.

Some of those biomolecules did persist in a relatively intact state. That much was clear from the organic objects within the peat – they had to be made of something. It was also already clear that the less conspicuously organic remains, such as pieces of pottery, retained some biomolecules. As early as the 1930s, a Boston scientist, Lyle Boyd, realized that the kind of antibodies that could attach themselves to blood proteins found throughout living tissue would also attach themselves to tissue taken from mummified bodies, and she went on to check the blood types of several hundred ancient Egyptians. By the 1970s various analytical chemists, such as Rolf Rottländer in Tübingen and John Evans in London, realized that methods of analysis in organic chemistry had reached levels of sensitivity that would allow slight traces of fatty/oily substances or 'lipids' to be detected inside ancient pots. They went on to use infra-red spectroscopy to track down the animal fats, plant oils, and even cooked eggs that once occupied some of the ancient pots unearthed by archaeologists.

Among the various molecules of which life is composed, we would anticipate the best survivors to be these 'lipids'. The word is a generic term for organic substances that resist mixing with water, including fats, oils and waxes. Water is so important to disaggregation and decay below ground that failure to mix with water is bound to confer resilience. But lipids are not all that survives. During the 1980s, two developments were leading us to believe that a far wider range of biomolecules might be isolated from ancient deposits. The first of these developments was a change in palaeontology, the study of fossils. Like archaeology, it had started out by giving prominence to the most durable and visible of finds, such as the rock-solid silicified shells and bones chipped away from their matrix with a geological hammer.

Through time, awareness grew of the survival of much softer tissues, such as in the remarkably preserved soft-bodied specimens from the Burgess Shale over which Stephen Jay Gould enthused in his book *Wonderful Life*.

The second development involved much more recent fossils. During the early 1980s, two publications appeared, one involving an extinct zebra-like animal, the other an ancient Chinese corpse. In each case, researchers claimed to have detected fragments of ancient DNA, the molecule in which life was encoded. Alongside lipids and proteins, DNA could also be identified in specimens of archaeological age. With the isolation of the molecule central to life's function, archaeology turned an important corner. It was difficult to put boundaries around the implications of recovering it from the past. The constraints on examining a living prehistory seemed to be falling away.

One of the first outcomes of these remarkable discoveries was that completely new bridges were hastily built between academic disciplines that had not hitherto had much to say to each other. Contact opened up between archaeologists, palaeontologists, molecular biologists, geochemists – specialists in widely different fields, who were beginning to sense a common interest. One of those meetings was between Terry Brown, a molecular scientist, Keri Brown, a prehistorian, Geoff Eglinton, an organic geochemist, and myself, by then a fully fledged bio-archaeologist. Born out of that meeting was a programme that the UK's Natural Environment Research Council put in place, in which 50 researchers around Britain put their minds to the problems, and their research efforts to solving them. For five years, the 'Ancient Biomolecules Initiative' found itself at the heart of a world-wide movement. Researchers in countless disciplines and countries became engaged in a molecule hunt that has, bit by bit, transformed our understanding of our own prehistory.

Looking back over three decades to my introduction to field archaeology, I can see that the change in our perception of what awaits discovery beneath our feet has been considerable. Those assiduously scrubbed pot fragments around which the whole exercise then revolved are seen now as the mere tip of a vast information 'iceberg'. Lower down on the 'iceberg' was a vast residue of the living organic world that ancient people experienced around them, indeed of which they were a part. It was a messy residue, browned, fragmented and falling apart, but it was definitely there, and in no small quantity. What could be gleaned from this prolific organic database? Back in the 1960s, we had only fragmentary answers to that question. Gradually, over the last three decades, the various surviving elements of those past organic worlds have been dissected and understood. One by one, the surviving fragments of past living worlds have been identified. First the more visible elements – bones, teeth, seeds and wood – became subject to rigorous analysis. The microscope has supplemented these with the less visible remnants – pollen grains, starch and silica bodies from inside plant cells, and the hard parts of insects and other invertebrates. Finally, molecular science has taken us one step deeper into this record. We can look within these fragmentary items to the molecules of which they are composed, and which determined their form and their biology. Some of these molecules are remarkably durable, surviving as evidence of living tissue that has otherwise completely dispersed. Other molecules take us to the very heart of life's structure. At the core is the molecular blueprint of life, DNA. We now realize that fragments of this life-encoding molecule can survive on archaeological sites that are twenty times as old as that ancient village in Somerset.

Within a few years of biomolecular archaeology becoming a reality, many stories about the human past have been rewritten, and others, out of reach of the traditional evidence at our disposal, have been narrated for the first time. The main stimulus for this new swathe of stories has been a search for the one particular biomolecule to which all others ultimately owe their existence, DNA. In the following chapters, those new stories are recounted, after first exploring how that unusual search reached its goal.

2

the quest for ancient DNA

a revolution in the life sciences

The double helix is a little younger than I am. To be more precise, I was a toddler of two years old when James Watson and Francis Crick rushed into the Eagle Pub at Cambridge to celebrate their breakthrough in understanding the structure of the molecule at the heart of all life. They had established that DNA was made up of two entwined sugar-phosphate strands, a twisted ladder whose cross-bars were made up of 'base-pairs'. These bases were chemical units reaching out from each sugar unit in the strands, to pair up with their partner on the opposite strand. Four types of bases could be identified, in a variety of different permutations. Watson and Crick had unravelled a mechanism by which a chemical code could be passed on from cell to cell, forming the blueprint for all life on earth.

My only involvement in all this was as a prolific factory of this remarkable molecule myself, in much the same way as any growing organism. The reason I place myself in the story is to emphasize the relatively brief history of the profound revolution in the life sciences that followed, and the pace at which our understanding of the molecular basis of life has grown. At the time of writing, we have the knowledge and power to replicate in the lab the entire process of reproduction, to transfer genes between unrelated organisms and to recover genetic information from organisms that have been dead for thousands of years. By the time this book reaches you, new discoveries will have been made. Such is the pace of DNA research. By the time I had grown from a toddler to a schoolboy, the double helix was part of the standard biology curriculum. By then, Watson and Crick's central thesis had been fully corroborated, and the way in which their base-pair code operated had been unscrambled. We learnt how the various permutations of the four bases found along the DNA molecule – Adenine (A), Thymine (T), Guanine (G) and Cytosine (C) – were read on to shorter strands of a very similar molecule called RNA. If DNA was the master blueprint, then the working diagrams were formed of RNA. They were moved around the cell to the point where the proteins were fashioned. Each RNA strand copied bits of the master code using almost the same bases, but the place of thymine was taken by another base, uracil. These working diagrams moved to cellular workshops called 'ribosomes' where the code was read to build proteins. Some of these proteins were the structural materials of life itself, materials such as the collagen in bone, or the keratin in hair. Others, such as the haemoglobin in blood, and hormones such as insulin, performed the major chemical tasks of life. A still larger group comprises the enzymes, described by Francis Crick as the 'machine tools' of biological chemistry, the things that fashion and enable most of the subsequent chemical pathways in the living organism. In short, the determinate link between DNA, RNA and protein was at the heart of all living processes. In little more than a decade after the recognition of the DNA double helix, the key elements of the genetic blueprint for the chemistry of life had been mapped out.

By the time I was passing the Eagle Pub myself on the way to Cambridge lectures in natural science, we were catching the first glimpses of another momentous advance. I had learnt at school about the clarity of vision of molecular scientists of how life's processes unfolded. At that time, the scientists were still observers. By the 1970s they were taking one step further and intervening actively. What enabled this was the discovery of the molecular scissors that could chop up the DNA strand at particular points on the sequence. These 'restriction enzymes' could home in on a characteristic sequence of bases and break the chain at that point. So, for example, an enzyme called '*Eco*RI' cuts wherever it finds bases in the order GAATTC, and another called '*Sma*I' cuts wherever it encounters CCCGGG. With enzymes such as these that allowed the double helix to be dissected, it was becoming possible to work with targeted fragments of the DNA sequence, to separate them out, to establish the order of their bases and to move them from one organism to another. In the later 1950s and 1960s the chemical blueprint for life was being mapped out. With the 1970s came the beginnings of intervention into that process, and the basis for gene cloning and genetic engineering as well as the possibility of recovering fragments of this chemical process from the distant past.

cloning from the past

Gene cloning was developed for medical and agricultural purposes, to allow the manipulation of genes that were affecting either health or productivity in a negative way. Different enzymes were used to cut out a targeted stretch of DNA from one species and insert it into the DNA of a 'host', a separate species that was easy to work with in the laboratory, a laboratory mouse or a microbe. In the host it could be bulked up through normal growth and reproduction, modified and reinserted in either the original host or a completely new host. Quite incidentally, the procedures perfected for performing these tasks were just the ones needed to track down those few traces of ancient DNA that might be surviving in an archaeological specimen.

In the case of ancient tissue, it could not be taken for granted that any DNA would survive at all. Archaeological preservation that appeared to be excellent quite typically involved only a selection of the molecules within the original organism, with an expected bias towards the dense structural molecules rather than those doing sensitive biochemical work. If there was any DNA surviving, then it was more than likely that it would be damaged and fragmented, and present in very small quantities. Because the gene-cloning process starts with cutting the sequence at very specific points, the 'restriction sites', it would be possible to design the restriction enzymes quite close together so that even short fragments with some internal damage could be isolated. Within the host organism, the normal processes of growth and propagation would transform the tiny quantities into manageable amounts for study. It was only a matter of time before cloning would be attempted on ancient tissue.

The obvious tissues to begin with were those in which biological breakdown had at least been arrested by a shortage of either water or oxygen. Attention turned to a range of ancient 'mummies'. Mummified bodies are those which have desiccated so speedily after death that a large part of the soft tissue remains in place on the skeleton. The process can be natural, artificial or a combination of the two. In artificial mummification, the internal organs in which breakdown begins are removed, and a variety of substances added to fix the soft tissue in various ways. One good reason to be optimistic about molecular survival was the appearance of mummified soft tissue under the microscope. As far back as 1911, attempts had been made to rehydrate the soft tissue from mummified bodies, and on a number of occasions the cell nuclei could be made out in these reconstituted tissues. If the nuclei were visible, perhaps their key components could still be found.

In 1981, the first claim appeared in print. Two Chinese scientists, G. Wang and C. Lu, isolated and identified nucleic acids from the preserved liver of a corpse from a 2,000-year-old Han dynasty tomb from Ch'ang-sha, the capital of Hunan Province. The result was published in Chinese in a journal not widely available in the West. The find made little impact until an American group set out on a similar quest. The first focus for these new ideas was the University of California at Berkeley. Around the time of Wang and Lu's publication, a number of scientists from around the Berkeley campus were speculating about ancient DNA. These scientists met together as the Extinct DNA Study Group, and tossed around ideas about where the science might lead. One Berkeley scientist, the geneticist Allan Wilson, had taken on a graduate student to track down ancient DNA from museum specimens. In many ways a preserved museum animal is very similar to a mummified body. Its preservation has involved rapid drying, removal of some internal organs and the addition of certain preservatives. Wilson's student, Russell Higuchi, was interested in a species that had been sighted by a number of eighteenth- and nineteenth-century travellers to South Africa. Charles Darwin made mention of the 'quagga' in his journal as the Beagle passed beneath the Cape of Good Hope in the 1830s. It was a timid, zebra-like animal, distinguished from the latter by the restriction of its stripes to the front of its body. Half a century after the Beagle had sailed, quaggas were extinct, surviving only as museum specimens. One such specimen was held by the Museum of Natural History at Mainz in Germany.

The Mainz specimen was not just a skeleton; some dried skin was also attached. Higuchi took a sample of this skin and observed some dry muscle tissue attached to the underside. He sampled this muscle tissue, cleaned it up, and separated its molecular components into fractions. Then came the critical test. The fraction expected to contain any surviving DNA was mixed with DNA from a modern mountain zebra. In theory, quagga DNA should be sufficiently similar to zebra DNA for the two to bind. If any quagga DNA did survive, it should bind with the zebra DNA. Higuchi came up with a positive result.

Having found quagga DNA, he could then go on to isolate it and sequence the bases that made up its genetic code. Two lengths, each of just over 100 base-pairs, were identified and positioned along the genetic sequence of the living horse genus *Equus*. One length belonged to a gene of unknown function, the other to a gene used in the energy management system of the animal's cells. Having established what the strands of ancient DNA were, the slight differences between those sequences and the corresponding sequences in other mammals could be used to build a 'family tree' of species ancestry, or 'phylogenetic tree', to use the correct term. The tree placed the extinct quagga very close to the zebra, somewhat further from the cow and even further from humans. A century earlier, the genus *Equus* had been used as a key example of how fossils told the story of Darwinian evolution. Now the same genus had come to the fore again, with the quagga as the first extinct animal to reveal a small part of its genetic code.

The quagga had long since ceased to be. Its bones no longer carried blood and its tanned dried skin no longer protected its body. Yet the very same information that brought it to life in the first place was, in part, recoverable. Its DNA had come alive again – the very process of cloning demanded that it do so. The bacterially carried strand was as much a part of a living cell as the bacterial DNA itself. Another way of looking at this strange phenomenon is that, by targeting and locating ancient DNA in a dead organism, the analysis moves beyond the distinction between life and death to the encoded information upon which that distinction is constructed. Among all molecules, ancient DNA probes more deeply into past life than any other. If Higuchi could reach his target with a well-preserved quagga, what about a well-preserved human, such as the Chinese scientists had studied?

About the same time that Wilson and Higuchi were looking back in time from the vantage point of molecular biology, a young Swedish Egyptologist, familiar with much older specimens than quagga, was looking forward in time to the newly emerging possibilities of molecular science. Having progressed from his Egyptology studies to a Ph.D. in molecular medicine, Svante Pääbo won a grant from the University of Uppsala Faculty of Medicine to look into the possibility of DNA surviving in ancient humans. He gathered samples from Egyptian mummies held in various European Museums, some of which dated back to the third millennium BC. A piece of skin was carefully taken from a mummified woman's ear and a stain called ethydium bromide was applied. This stain attaches itself to DNA, and DNA alone, and is used to mark out visually its presence or absence. When Pääbo looked down his microscope he could detect nuclei within the skin cells of that ear, nuclei that carried the stain that demonstrated that her DNA had survived. This he managed to repeat with other mummies, which spurred him on to attempt to clone the DNA. He eventually achieved this with the skin of a young boy, isolating a 500-base-pair sequence containing a so-called '*Alu*-sequence'.

This ability to reach back in time and recover the molecule at the very heart of life was attracting increasing interest, but the discovery would remain a novelty unless there was real information that could be gleaned from it. When Pääbo published his findings of DNA from the Egyptian mummies in *Nature*, he speculated optimistically about the possibilities of the new research (Pääbo, 1985). Perhaps we could start looking at gene frequencies; virus evolution; descent of the Nile valley population; Pharaonic and interfamily relations. It was one thing raising these possibilities, but Pääbo knew that he had to find some real information from ancient DNA to maintain the interest his work had attracted. Around this time, a further step forward in methodology helped him significantly on his way.

mimicking nature: the polymerase chain reaction

One of the challenges of ancient DNA analysis was to make sense of tiny quantities of genetic material. This is a challenge that living organisms meet with every new generation. Each adult individual is only formed as a consequence of the cell system's ability to make sense of two individual sets of genetic code, one from the egg, the other from the sperm. The natural process of doubling up at the heart of cell proliferation makes it possible to work with DNA information that starts in tiny quantities. The double helix unwinds, each strand forming the template to construct a new partner. Where one double helix once was, now there are two, then four, then eight and so on. The gene-cloning methods that Pääbo had adopted used a fast-growing organism to perform that doubling up. It was not long before someone realized that we could go one step further and automate the process.

Our growing understanding of the DNA molecule was spawning an entirely new profession, that of the now familiar 'genetic engineer'. One of the first generation of these molecular mechanics was Kary Mullis, who was as familiar with the curves and bends in the double helix as a car mechanic might be with engines and carburettors. Beyond this, he was a lateral thinker *extraordinaire*, whose achievements included the invention of a light-sensitive plastic, and the publication in *Nature* of a consideration of the cosmological consequences of time reversal. The next major step in DNA manipulation he worked out, not in a laboratory surrounded by equipment and reagents, but in his head while on a weekend drive through the Californian redwoods – and, incidentally, while chewing over a problem that was quite different from the one he was about to address with astounding success.

Soon after the structure of the double helix itself had been established, a key enzyme in the process, called polymerase, was identified. This enzyme built new double helices, and repaired slightly damaged ones, by lining up the links in the chain, the nucleotides, in the right order. It was in action more or less continuously in the growing organism, taking care of DNA replication and repair. Kary Mullis was toying with how he could use the enzyme to establish the sequence of any DNA chain. He played with the molecule in his mind, imagining how a raised temperature would decouple and unpeel the two strands of the double helix. He then thought of lowering the temperature so that the strands once again annealed, but not with each other. They annealed instead to added 'primers' – very short strings of nucleotide that bound with particular short stretches of the separated DNA strands. These primers attached themselves to either end of the stretch of DNA, one on one strand of the disaggregated helix, one on the other. They acted something like bookends, within which the polymerase enzyme could stack the nucleotides in the right order. Pretty soon he had two stretches of DNA, where before there had been one, but corresponding only to the section bounded by the bookends, the primers. It was a question of controlling the temperature, adding primers, and then polymerase and enough free nucleotides for the enzyme to set to work.

At this point Mullis's imagination drifted from the sequencing problem to something else. If it was possible in this way to get from one DNA molecule to two, fairly quickly and outside a living cell, the cycle could be repeated to get from two to four, four to eight, and so on. With the right sort of equipment, it should be possible to double up again and again, and in the space of an afternoon turn one single DNA molecule into a billion identical molecules. In a lateral thinker's head on a moonlit California drive, the Polymerase Chain Reaction was born.

The PCR, as it is known, fortunately transferred well from imagination to reality. It is now a key molecular tool that has transformed all aspects of DNA research, not least the quest for ancient DNA.

making sense of the fragments: PCR in action

Svante Pääbo was one of those who quickly picked up on PCR and applied it to ancient DNA. Armed with the new technique, he managed to move forward from simply recognizing the DNA to working with informative sequences. He had by now joined Allan Wilson's lab in Berkeley and was exploring the possibilities of North American material. A group of archaeologists from Florida State University had been digging in a swampy pond in Brevard County, Central Florida.

The Windover Pond is a flooded sinkhole, a valuable source of drinking water for humans and animals alike. Humans had used such sinkholes ever since they first spread across the New World, and that is why sinkholes excite the interest of archaeologists. A sinkhole such as this one could contain any number of remnants of people's visits to quench their thirst, in sediments accumulating over thousands of years. As the archaeologists would discover, it could also contain the human bodies themselves. Beneath the shallow water that remains in the sinkhole lay several metres of peat that had built up during 12,000 years of visits. As the archaeologists dug down into the peat, they found the remains of a middle-aged woman about three metres down. There are quite a few bog bodies in the archaeological record, but one particular thing was striking about her. Within her skull she had what looked like an intact brain.

This remains a most unusual find. The brain was one of the organs the ancient Egyptians removed, as it was subject to autolysis, the self-triggering breakdown that initiates bodily decay. The soft tissues that most readily survive in the absence of water or air are skin, hair, gut and tendon. Muscle is not infrequently found, but brain is an unusual survival, although not confined to this woman alone. As the team dug further, another thirty-nine bodies came to light. Eight of these also had brain-like material within them, which the team was now in a position to subject to special scrutiny. Each excavated skull was subjected to a range of non-invasive scanning procedures, prior to close examination under a microscope. It transpired that the actual brains lay within the soft mass, and had shrunk to a quarter of their original size. The shrinking of these soft, granular, tan-grey organs had not, however, obscured what they were. Within the brain could be detected thalamus, basal nerve ganglia and the ventricular system. Under a microscope, individual cells could be made out, and through ethydium bromide staining the remains of ancient DNA visualized. Its concentration was estimated at about one part per million, that is 1 per cent of the original concentration of DNA in the living brain.

Pääbo was struck by the exceptional preservation within these Florida sinkholes. He made contact with the Miami anthropologist John Gifford in order to get hold of a similar brain from another sinkhole, at Little Salt Spring. Working with his 7,000-year-old sample, he amplified and sequenced enough of the DNA to recognize not only that it was human DNA, but also that it was of a particular genetic lineage not hitherto encountered in the New World. Here was clear evidence that the uses of ancient DNA could extend beyond recognition to the uncovering of new genetic information, not available from living individuals.

Over the following four years, Svante Pääbo and Allan Wilson applied the Polymerase Chain Reaction to a new set of ancient tissues, including Egyptian mummies, two extinct mammals, a marsupial wolf and a giant ground sloth, and, in addition, to a relatively young piece of dried pork. The last two yielded what were at the time the oldest and youngest specimens of ancient DNA. The ground sloth was 13,000 years old, and the dried pork four years old. There was a range of dates in between: the 140-year-old quagga, mummified humans between 300 and 4,000 years old, and an 8,000-year-old brain. The apparent state of survival followed no clear pattern. The youngest and the oldest, the sloth and the pork, were in much the same state of disrepair. Such DNA as survived was broken up into very short lengths, often of no more than 100 or so base-pairs. All the damage was apparently done in the first few years before full desiccation had taken effect. A very ancient body desiccated within days or weeks might retain more intact DNA than a very recent body dried over a period of months or years. In general, the time taken for even a rapid dehydration of something the size of a human body was sufficient to reduce ancient DNA to very short fragments, whatever its age. The lengths of ancient DNA that survive suggest that we can expect length fragmentations of the source DNA by a factor between a hundred and a million. The analyst of ancient DNA is looking at small, individual pieces of a very, very large jigsaw, with at most a partial glimpse of the lid of the box.

Pääbo and Wilson's work with PCR also cast some light on why the cloning had sometimes presented a different picture. What they noticed was that, although the PCR itself indicated very small quantities of highly fragmented DNA, the overall survival of DNA in dry tissue was significantly greater. There was some reason why a lot of the DNA was not responding to the PCR. When Pääbo looked more closely at the material, he noted a marked contrast in the pattern of bases that formed the cross-links in the double helix. The bases fall into two types. The 'pyrimidines' have a single ring structure, and include cytosine, thymine, and the latter's RNA counterpart, uracil. The 'purines' have a double ring structure and include the remaining two bases, adenine and guanine. In Pääbo's tissues, the pyrimidines had fared far worse than the purines. Judging from how much had survived, the pyrimidines were at least twenty times more susceptible to damage in mummified tissue than the purines. As we shall see in due course, different patterns prevail in different environments. What their comparisons between PCR and cloning led the researchers to suspect was that the latter was doing rather more than just amplifying the ancient DNA. A living organism has within its molecular toolkit a series of repair mechanisms. To stay alive, it has to maintain vast lengths of DNA sequence in good condition, and the living cell is continually repairing minor sequence flaws. What Pääbo concluded was that the host used to clone the damaged ancient DNA was doing something similar and creating cloning artefacts that gave a rather better impression of DNA survival than was actually the case. By contrast, the use of PCR had trimmed the amplification process down to its raw chemical mechanics. It was much more likely to provide an accurate reflection of what survived. Even with this approach, there are amplification artefacts and certain simple forms of repair, but PCR both gave a realistic reflection of ancient DNA, and pushed the sample size limits to their extreme. It was, in principle at least, possible to amplify a single surviving DNA molecule to quantities large enough to study. This enormous potential opened up the vistas of analysis, and the aspirations of a growing band of ancient DNA scientists. By the end of the 1980s the race was on for the oldest DNA, and journals such as *Nature* and *Science* were poised at the finishing line.

into the distant past

Things moved very fast. By 1990, attention had turned to the unmistakable outline of a leaf, darkened by long-term burial but retaining much of its original physical form. Botanists had given this extinct species the name *Magnolia latahensis*. The leaf came from a basin in Idaho popular with bikers and holiday-makers, at a place called Clarkia. The basin is made up of deposits of clay and ash that rapidly accumulated within an ancient lake 17–20 million years ago. The Clarkia site had been visited by fossil hunters for years, and plant tissue and fish remains found there had been the subjects of much study. Karl Niklas had looked at leaves from the site a few years earlier. He demonstrated that cell structures were discernible, such as the chloroplasts, the tiny green bodies within the cell that capture light and turn it into biochemical energy for the plant's use. He also found that certain of the molecules that had endowed the leaf with its original colour remained intact. Now Edward Golenberg at Detroit, Michigan, went one step further and detected within those chloroplasts the remains of their DNA, thousands of times more ancient than any DNA Pääbo and Wilson had amplified.

These leaves were compression fossils, caught up and compacted after some change in the drainage pattern accelerated the lake's infill. Their excellent preservation resulted from the absence of oxygen at the lake bottom, an extreme version of what led to the preservation of peat discussed in the previous chapter. It was oxygen that was damaging the pyrimidine bases in the mummified specimens, and its absence from the Clarkia leaves might explain why, as well as being older by far than other ancient DNA specimens, the amplified sequence was also longer. The chloroplast DNA sequence was 790 base-pairs long. This *Magnolia* leaf had raised the stakes for ancient DNA. Clarkia in Idaho became a magnet for the newly emerging band of molecule hunters.

Among these were the members of the Extinct DNA Study Group. These Berkeley palaeontologists and biologists had for some time been interested in tracking down DNA that was millions, rather than thousands, of years old. From the early 1980s they had been toying with the idea of a quite different source of ancient DNA than the mummified tissues so far examined. This source was to be found on the islands of the Dominican Republic where, 30 million years ago, lagoons were fringed by a ring of now extinct resinous trees. Beneath the canopy of the trees, a group of bees engaged in the sometimes dangerous task of gathering the sticky fresh resin that exuded from the trees. From time to time, these diminutive resin-collectors would get caught up in the object of their desire, and their struggle to escape would only drag them further into their viscous tomb. Millions of years later, the resin had fossilized to amber, but the whole body, and even the delicate wings, had been captured and frozen in time. Not only had the amber matrix become a permanent prison, but it also formed an effective barrier against oxygen and other gases, and served as a mummifying agent. Among its constituents are sugars and compounds called terpenes which contribute to preservation in two ways. They dehydrate the entrapped specimen, and inhibit microbial breakdown. The Extinct DNA Study Group had their eye on this seemingly ideal molecular time-capsule. Among the group was the entomologist George Poinar Jnr. His early speculations about these insects in amber inspired the author Michael Crichton to start work on *Jurassic Park*. A few years later, and spurred on by the *Magnolia* result, George Poinar worked with his son Hendrik and another California scientist, Raul Cano, to give substance to those speculations.

After the 17–20-million-year-old fossil leaf, it was only a question of time before the Berkeley group managed to amplify the bee's DNA. Others were on the same track. At the American Museum of Natural History in New York, Rob de Salle and David Grimaldi were looking at termites from similar Dominican amber. By 1992, both groups had succeeded in pushing back the age of the oldest DNA to 30 million years. By 1993, Hendrik Poinar and Raul Cano would have pushed the time range even further back, with an amber-embedded weevil, to 130 million years old. The time constraints were falling away. Brian Farrell at Colorado had a preliminary positive result from a 200-million-year-old fish, and Noreen Tuross from the Smithsonian Institute was attempting to amplify DNA from a 400-million-year-old brachiopod. It seemed that life's fundamental code could be recalled from the depths of geological time, and if that code could be recalled, whole living worlds could be contacted and viewed in meticulous detail. But all was not plain sailing – there were clouds on the horizon.

a time for caution

Back in 1990, the Extinct DNA Study Group had not been the only ones to follow in Edward Golenberg's footsteps. Up until the *Magnolia* publication, the front runner in the race for ancient DNA was emerging as Svante Pääbo, the Egyptologist-turned-molecular-biologist, now with Allan Wilson's renowned lab in Berkeley. His sequences from Egyptian mummies and pickled brain tissue had caught everyone's imagination, and he was enjoying the limelight in this novel and exciting field. However, as he travelled to Idaho in August of 1990, it was with the knowledge that there was a new kid on the block, whose DNA targets were thousands of times older than his. Pääbo needed to have a look for himself, and he did so with a certain amount of scepticism.

His scepticism had a sound scientific basis, arising from DNA research in a quite separate field. He had read the papers of a cancer research scientist working in London. Tomas Lindahl's research initially had nothing to do with fossils. He was concerned with the stability of DNA, a topic of great importance for understanding human health, for example in relation to ageing and cancer. A decade before anyone had thought of extracting ancient DNA, he had published estimates of the rates of breakdown of various parts of the DNA molecule. A process to which he paid particular attention was that of 'depurination'. The purines were the two DNA bases, adenine and guanine, that proved to be more resilient to decay than their pyrimidine counterparts in the presence of oxygen but without water. In contact with water, however, the purine residues break down, and this process of depurination also weakens the sugar phosphate backbone to which the bases are attached. Working from Lindahl's estimates for the rate of depurination, Pääbo was able to gauge how long a fragment of ancient DNA might survive when in contact with water. The sequence amplified by Golenberg was long by ancient DNA standards, measuring 790 base-pairs. Pääbo estimated that the survival of such a length could be measured, not in millions of years, but in thousands. From the consideration of chemical kinetics alone, the strand amplified from the 17–20-million-year-old fossil was at most 8,000 years old. Pääbo took his own samples for study in the Berkeley lab, and by early 1991 was ready to challenge Golenberg's findings.

In March 1991 the two came together to put their cases at the meeting rooms of London's Royal Society. The audience was made up of the diverse collection of researchers who had been drawn into the molecule hunt. There were geologists, archaeologists, chemists, geneticists, taxonomists and many others. Among them were the senior scholars who had laid the foundations of the new field. They listened as the two young scientists stood up in turn to make their cases.

Pääbo and Golenberg shared a rather attractive aura of enthusiasm common to researchers on the cusp of some new scientific departure, a discreet discomfort with the lapel mike, laser pointer and conference clothes, and an urge to get back to the field or the lab, where the real action was. Pääbo however had gently moved on from this position. He was no longer simply the bright young star of the field, but was getting used to a new role as traffic policeman in a convoy moving with rather too much momentum for its own safety. His counterpart was aware of this; he had already read a 'cautionary note' placed in a recent issue of *Current Biology* by Pääbo and Wilson. In a slightly nervous presentation, Golenberg acknowledged the need for care and control, but also pointed out there were now several drivers in the fast lane. From the same ancient Clarkia beds, researchers at Washington State University claimed to have sequenced DNA from a *Taxodium* leaf. An Idaho group had reported amplifying a sequence from oak of the same age. Then of course there were the amber specimens that went even further back.

Pääbo was not to be deflected. He reiterated the kinetic arguments drawn from Lindahl's work, and outlined with care his own group's attempts to replicate *Magnolia* DNA. They had found no plant DNA within the leaf, but bacterial DNA was present. The inference was that Golenberg's result arose from contamination. This was something of a turning point in the quest for ancient DNA. The excitement about new results had infected everyone, including the referees used by high-profile scientific journals. There was never any doubt that a method like PCR, that could theoretically find a single surviving molecule, was extremely vulnerable to contamination, even in the cleanest of labs. However interesting and sincere Golenberg's arguments were, there was a palpable sense that the convoy of ancient DNA hunters should adopt a more cautious respect for the highway.

That more cautious approach came naturally to the meetings' organizer, Geoff Eglinton. He was an established figure in the science community and had no need to race for the pages of *Nature* and *Science*. He also had a lifetime's experience of the problems of research into biomolecules, having looked for them everywhere from the oldest rocks in the ground to the first samples of moon dust brought back with the Apollo mission. This experience gave him the sense that the quest for ancient DNA was proceeding in a somewhat myopic manner. Different DNA scientists were asserting different things, and there was much discussion of whether the results of this or that lab could be believed or not, rather than how best their results could be explained. The key to scrutinizing the results was to broaden the analysis from DNA to other molecules. The reason something as delicate as a leaf survives at all, even during its lifetime, is because it wears a tough waterproof coat. Its waxy cuticle contains molecules that are among the most persistent in nature. If there is exceptional survival of the less durable molecules such as DNA within the leaf, then the persistence of that waxy protective exterior should presumably reflect that.

Eglinton was another visitor to the Clarkia site, from which he returned with a sample of the darkened ancient leaves. Back in the lab, he took a scalpel to detach the outer layer of one of the fossil leaves from the Clarkia bed, and used a specialist solvent to dissolve what lipids remained. This solution was subjected to a highly sensitive method of organic analysis called GC/MS (gas chromatography/mass spectrometry). What the analysis demonstrated was that the waterproofing lipid molecules in the Clarkia leaves were in reasonably good shape. There may indeed have been pockets of cellular tissue protected from water, in which case Lindahl's rate estimates need not apply. His studies had been on DNA in an aqueous medium. However, there was something else rather odd about the Clarkia lipids.

Eglinton did not stop there; he also wanted to contrast the lipid profile in what was visibly discernible as the leaf's waterproof envelope, and the lipid content elsewhere in the deposit. What one might expect is a sharp contrast between the two, confirming that the lipid signal received corresponded to the envelope itself. Instead, what he found was that the leaf lipids were not confined to the leaf itself. They had dispersed into the surrounding sediment. Rather like the Somerset Levels peats, the exceptional appearance of the leaves, buried so long ago under the Clarkia muds, betrays how speedily so many decay processes were arrested. However, just as in the more recent peat, they weren't arrested completely. The Clarkia leaves changed colour, and we can guess from peat processes how that happened. The lipids also intermingled with the sediments, and we are less clear what was going on there. At some stage, living bacteria came on to the scene, as is clear from Pääbo's results. When they came is less clear. They might represent a quite recent contamination. Could they, on the other hand, not have lived within the sediments for much longer, and could this be linked to the reworking of the lipids? These are questions that remain open.

Scientific funding is fickle. As with economic predictions, it depends as much on 'community confidence' as on cool rationality. There were undoubtedly questions left unanswered by the Clarkia DNA papers, but the answers would variously be negative, cautious or complex, and perhaps lack the immediate impact a future front page of *Nature* would require. There was a sense of a downturn in the prospects for *Magnolia* DNA, and the science funding bodies chose to invest elsewhere. In Michigan, Edward Golenberg shifted of necessity to new areas of research without having gained the means to resolve the *Magnolia* issue once and for all.

Back in England, some of us were having more success in persuading the UK research councils to follow this avenue of research. We were keen to shift the agenda away from the race for headlines that had begun to characterize the quest for ancient DNA, to support research into the dynamics of the various biomolecules involved, and to look more carefully at what they could tell us about the past. It was another eighteen months before the Ancient Biomolecules Initiative (ABI) was under way, funding projects in a number of different universities and institutions around the country, several of which were directing their attentions to ancient DNA.

getting the science right

One of the first issues the ABI programme needed to tackle was that of the DNA from deposits far too old to tally with Lindahl's rates of DNA breakdown. By this time, Tomas Lindahl had addressed the issue of ancient DNA in his own publications. Now that he had entered the debate, one of the critical variables to emerge was water. One might think that water was quite an easy thing to assess. Either an object has remained wet or dry, and that is that. Things are, however, rather more complicated, largely because biological structures are more complicated. As Eglinton's work reminded us, they contain lipid barriers that stop water moving around, though it is very difficult to establish whether such barriers have remained intact – especially the microscopic barriers around cells and within tissue. Water's impact also depends on what is dissolved in it. Anyone who has lived by the sea and seen the relative corrosive impacts of salt water and rainwater will know that. The situation is yet more complex underground where dissolved minerals can get mopped up from the soil water by complex organic molecules. In sum, just because we can see that an ancient object is surrounded by wet sediment, as in the case of the Clarkia leaves, and indeed of the Windover bodies, it does not follow that all the contained biomolecules have been exposed to the effects of water.

An alternative approach was developed by Hendrik Poinar, by now the front runner in the 'oldest DNA race', thanks to his 130-million-year-old weevil. With great foresight, Hendrik crossed the board to join the sceptics in Svante Pääbo's new lab at Munich. Here, he developed an elegantly simple approach to the problem of water. It is difficult to establish whether water has successfully reached the inner recesses of a complex organic object, so rather than look for the water itself to discover whether it had reached the DNA, we could look instead at a neighbouring molecule that was also affected by water, to see whether water had left its imprint. Poinar turned his attention to the proteins surviving alongside the DNA.

The amino acids that make up any protein chain have a rather interesting property. As with any molecule, it is possible to make a three-dimensional model of an amino acid, displaying its atoms and the bonds between them as a kind of geometrical construct. In the case of amino acids, a mirror image can be taken to produce the model in reverse. So two different geometrical models of an amino acid share the same atoms, and the same links between them, but each is the mirror image of the other. For convenience, we can refer to these as the left-handed and right-handed forms of the amino acid. If we artificially prepare amino acids in the lab, the two forms occur in equal amounts. One of the remarkable, and indeed little understood, features of the living cell is that one of the two forms, the left-handed form, predominates. It is only after death that the balance gradually moves back to the more stable, even proportions of left- and right-handed forms. This happens by the amino acids spontaneously 'flipping over' or 'racemizing' at a rate that can be measured, but only in the presence of water. If the proteins retained contact with water after death, we would expect racemization to proceed. Otherwise, the marked imbalance between left- and right-handed forms of amino acid would persist.

In amino acid racemization, Hendrik Poinar found a test to assess the internal dryness of an ancient biomolecule. It was not just a measure of how dry the specimens were now, but constituted a kind of cumulative record of any tiny quantities of water that might have reached the biomolecules in the past. If any of the geological specimens really did contain ancient DNA, we would expect the molecules to have been sufficiently protected from water for their neighbouring amino acids to show minimal signs of racemization. Poinar assessed and disposed of various of the geological samples, but one interesting result came from the amber-embedded insects from which he, with his father and Raul Cano, had amplified DNA. These specimens displayed very low levels of racemization. Here, perhaps, was one place where their geological DNA findings would stand up to close scrutiny. The terpenes within these amber prisons might have kept the water sufficiently at bay for at least a small element of Jurassic Park to shift from fiction to factual possibility. At the first workshop of the Ancient Biomolecules Initiative, Tomas Lindahl agreed this was the right place to look for the longest surviving DNA. Andrew Smith and colleagues at the London Natural History Museum set to work scrutinizing this possibility.

They wanted to give themselves the best possible chance of encountering the DNA locked up in amber. So they tried five different DNA extraction techniques, targeted a variety of gene fragments that were suitably short (down to 110 base-pairs in length) and looked at several parts of the entombed insect's body. Their first attempt was made on those extinct resin-collecting bees from the Island of Dominica, from which Cano and De Salle had recovered DNA. The amber chosen showed no signs of cracking, and Hendrik Poinar's amino acid racemization tests had shown the entombed insect tissue to be free of water. The museum team came up with a negative result. There seemed to be no amplifiable DNA within the amber. They needed to look further. Perhaps a smaller body would be buried more quickly and dry out more effectively, so they expanded the range of species to include smaller insects. It would be interesting to see whether younger samples were any different, and bees from African copal, a fossil resin less than 100,000 years old, were assessed. None of these satisfied the team that any insect DNA remained.

This is not to say they amplified no DNA. In fact, of the 156 trials they made, seven samples produced PCR products of the expected number of base-pairs in length, but on closer scrutiny these revealed more about how easily the minutely sensitive technique of PCR could deceive even the most rigorous scientist. Although they were the right lengths, sequencing of base-pairs showed them to be artefacts of the method. The most interesting of these artefacts was a length of DNA built entirely from copying the short sequences on the primers, over and over again. This accounted for two of the seven. The others clearly had some input from mammals or fungi, showing how sensitive PCR is to contamination, even in very clean labs. None of the seven sequences came close to an appropriate insect sequence.

After three years of careful work, replicable amplification eluded the team. Furthermore, it eluded other labs in America and Switzerland, who had set about addressing the same issue. Everything pointed to a negative result. To put the cap on it, another member of the ABI team, Derek Briggs at Bristol, looked at the survival of what should be far more durable molecules than DNA, molecules such as chitin and certain structural proteins. Even these were broken down within the specimen. Seeing them with the naked eye, and especially with the help of low-power magnification, we may admire the beauty and apparently intact state of these amber-embedded insects, seemingly as fresh as the day they undertook their task of collecting resin with a little too much gusto. Their form seems wonderfully intact, but their molecules are not.

This was by now becoming a recurrent message of biomolecular analysis. In the early days of seeking out biomolecules, the visual state of preservation was the only clue to molecular preservation. If the cellular structure was visible, and especially if the nuclei were visible, then surely some of the molecules had survived. It remains a useful guide at the broader level, but not an invariable indicator. An increasing number of cases of seemingly intact tissue with greatly transformed molecules were coming to light. Geoff Eglinton had found that the content of the Clarkia leaves had migrated out of the leaf tissue into the surrounding sediment. The ABI team were revealing considerable breakdown in amber-embedded insects.

from the very old and unusual to the commonplace

This first episode in ancient DNA analysis moved between a series of rather unusual objects, with what has proved to be a flawed foray into older and older specimens. It progressed from the liver of an ancient Chinese corpse, to the skin of an extinct quagga, on to Egyptian mummies, and from there to some extraordinary brains recovered from Florida sinkholes. We then heard of compressed Miocene leaves and insects in amber. The assembly of sources was strangely reminiscent of the objects and curiosities with which the eighteenth century antiquaries would surround themselves.

Around the time that Svante Pääbo was producing exciting results from the preserved brains in the Windover Bog, members of a research group in Oxford, England, were turning their attention to a rather more commonplace archaeological find than any of the above. As ever, the team brought together researchers who had moved in interesting ways across the boundaries of different disciplines. Bryan Sykes was a medical geneticist; Robert Hedges was a physicist turned archaeological scientist, who had been at the forefront of Britain's carbon dating and was very familiar with the chemistry of archaeological materials, particularly bone; Erika Hagelberg had a Ph.D. in biochemistry but, wanting to expand her horizons, moved to the history and philosophy of science, and now hoped to bring together science and history in an altogether novel way. Having heard about these exciting advances with unusual archaeological specimens, the team decided to see what could be done with something very ordinary in archaeological terms, in circumstances where no soft tissue survived. They sought to find ancient DNA from archaeological bones.

Hedges gathered together a range of fairly typical archaeological bones, some just 300 years old, from an English Civil War cemetery, some first-millennium-AD bones from an Anglo-Saxon cemetery, a child's bones from a late prehistoric hillfort, and a 7,500-year-old human bone sample from a Judaean cave. Under Sykes's guidance, Hagelberg set to work, grinding up bone samples and preparing them for PCR. By the end of 1988, she had succeeded. From these samples, a gramme of powdered bone yielded up to five microgrammes of ancient DNA. The analysis had shifted from the unusual finds and museum acquisitions to the commonplace material of archaeological excavations.

In order to answer specific questions bone samples could be collected from a vast range of sites. Erika Hagelberg went on to amplify DNA from a mammoth tooth, deposited in the Siberian soils around 50,000 years ago. It was both a replicable result and consistent with Lindahl's dynamics. Shortly after Hagelberg had her first success with bone, Terry Brown and I met up and discussed the equivalent commonplace plant tissue from archaeological sites. Terry was a crop plant molecular biologist whose marriage to a prehistorian had aroused his interest in the past. My research had been in early agriculture, based on the plant tissue that is encountered on numerous excavations. I selected for Terry a series of the archaeological crop remains from different but commonplace conditions of preservation, and Terry got positive results within each category. At the same time that the very ancient DNA assays going back millions of years were looking vulnerable, the last few thousand years was looking considerably more possible. No matter how well an ancient amber prison had been hermetically sealed, or how speedily a Miocene leaf had been compressed, the oldest DNA looked doubtful. However, within the relatively recent history of our own species, we did not have to confine our attentions to unusual samples. DNA was potentially extractable from the most commonplace organic materials on an archaeological site.

As the doors to geological DNA began to close, those to archaeological DNA were opening up. The next chapters examine the stories that are unfolding as we pass through those doors. But before we leave geological time behind, we can look at one further attempt to take the methodology back into areas that were still the preserve of fiction.

one last stab at jurassic park

In the years leading up to this point, life and art had been having some difficulty keeping pace with one another. In April 1990, *Nature* published Ed Golenberg's paper on Miocene *Magnolia*. Within a month, Michael Crichton's publishers had passed the galley proofs of his forthcoming novel to Universal Pictures. By the time Steven Spielberg had released the movie *Jurassic Park*, Crichton's source of inspiration, George Poinar Jnr, had worked with his son Hendrik and Raul Cano to publish a series of papers on amber-preserved DNA going back 130 million years. In London's Natural History Museum, the work that would sound the death knell of the amber story was soon to begin. However, Crichton had picked up the other possibility in his story, that the bones themselves might retain Jurassic period DNA.

Early in 1992, Mary Schweitzer, who happened to be a graduate student of one of the blockbuster film's palaeontological advisers, detected what seemed to be nucleated blood cells in a bone of the dinosaur *Tyrannosaurus rex*. The bone was 65 million years old. By the summer of the film's release, she believed she had detected DNA within the bone. Schweitzer was wisely cautious about her find and well aware of the problems of contamination.

In broad medical terms, the structure of bone was well understood. It was a living tissue in which the bio-mineral matrix was fully permeated with blood vessels. In other words, the outside world had plentiful access to a dead bone through its natural openings. Fungi and other decay organisms could make their way in and speedily digest most of the lingering DNA. However, PCR could work with tiny residues as well, so it was not just a question of the general survival. Some cells could become completely engulfed by bio-minerals, and their DNA isolated from water and biology even before they left the original body. So long as the bio-mineral remained intact, and engulfed at least some cells, then bone tissue would be a very effective protector for ancient DNA. This became increasingly evident as studies of ancient DNA from bone increased. Some of the best results came from *Moa* bones preserved in New Zealand peat bogs. The DNA within the bone was clearly being kept isolated from the watery peat. Perhaps that dry enclosure could also work for geological bone.

Schweitzer was not the only one with an eye open for DNA in fossil bone. A microbiologist from Utah who had been working on colon cancer decided to give it a go. Scott Woodward was impressed by various examples of excellent preservation within peat, and reasoned that very ancient peat eventually became coal. He had grown up in mining country and knew about the occasional dinosaur tracks and dinosaur bones that turned up within coal. In late 1992, he asked some of his friends in the coal-mining industry to look out for ancient bones. Soon after, they provided him with some large bone fragments from a sandstone band within coal measures believed to be 80 million years old. The bone was fragile, brittle and waxy, which seemed promising. It had not, apparently, been heavily replaced with minerals, as many fossils are. Under a microscope, cellular structures could be made out. Some of those structures picked up stains designed to attach preferentially to biomolecules. Woodward was optimistic.

He decided to look for a certain gene connected with the way cells manage their energy supply, the so-called cytochrome b gene. These can be found on the sub-cellular powerhouses known as mitochondria, which carry their own DNA independently from the cell nucleus. The cytochrome b gene is widespread among animals, and common to vertebrates, with slight variations between taxonomic groups. Dinosaurs should therefore have them. If he could pick up sections of the gene from his ancient bones, and compare them with living vertebrates, then he might have a basis for sticking his neck out further than the more sceptical Schweitzer. He designed PCR primers for the cytochrome b gene, and applied them to forty-two different extracts from the two bones. With many replications, the number of amplifications attempted came to nearly 3,000. Among these, the great majority were blank, but nine came up trumps.

The sequences he obtained were short, 174 base-pairs in length, but long enough to show a match with part of the cytochrome b sequence. Within this sequence there was a lot of variation, but variation with a certain amount of structure. In particular, some of that variation was attributable to the difference between the two bones. Other variation was due to the age and state of preservation of the biomolecules. Woodward concluded that the result was worth publishing as a possible example of ancient DNA from Cretaceous bone. The number of differences between his sequences and those of living animals pointed to a remote and extinct group, and the difference between the two bones implied that they had amplified DNA from not one species of dinosaur, but two.

Within six months of the publication in *Science* of Woodward's exciting find of 'dinosaur DNA', the same journal carried three pages of dense and carefully constructed critique of his argument. In that short time, twelve scientists from both sides of the Atlantic, including geneticists, palaeontologists, zoologists and virologists, had carried out new PCR amplifications and analyses, built new phylogenetic trees with the data, and lined up a battery of arguments for Woodward to address. Mary Schweitzer was one of those arguing that Woodward needed to take the phylogenetic analyses further. When she herself did so, she found that the dinosaur sequences had marked affinities with the human sequence. Svante Pääbo, from his new position as Professor of Zoology at Munich, was still keeping a keen eye on adventurous ancient DNA claims. His team decided to mount their search for dinosaur DNA not in amber or the coal measures. They chose to search for it in fresh human semen.

This seemingly eccentric quest did have a logic of its own. Sperm is an ideal medium for a separate examination of nuclear and non-nuclear DNA. The head of the sperm carries only nuclear DNA, whereas the tail needs those minuscule powerhouses within the cell, the mitochondria, to enable its movement. If a method can be devised to chop the head from the tail, then the separate examination can proceed. The gene targeted by Woodward was a gene on the mitochondrion, not the nucleus. It was this mitochondrial gene that was the basis of Woodward's analysis. However, one of the complicated things about DNA is that a particular sequence can crop up more than once within the same genome. There are a number of reasons for this to happen, some functional and others historical accidents. One example of the latter is that at some time in the distant past, copies of the cytochrome b gene transferred themselves from the mitochondrion to the cell nucleus. There, as inactive 'fossil' genes, they could embark on their own mutation pathway, unfettered by the constraints of natural selection. Pääbo's team suspected this might have something to do with Woodward's results, and so embarked on the sperm test to look at the nuclear and mitochondrial versions of the human genome separately. They used direct copies of Woodward's own primers to seek out the dinosaur DNA in the sperm fractions and in various other samples. They got a very good match between the Cretaceous dinosaur bones and the modern human sperm heads.

Their discussion put forward a number of ways of accounting for this unusual result. It may have been that a stray dinosaur had found its way into Pääbo's lab and contaminated the equipment in some way. A second possibility was that at some point in our ancestry hybridization with a dinosaur was involved. I am sorry to say that Pääbo's group favoured an explanation far less evocative than either of these. This third possibility was that human DNA carrying the confusing 'fossil' genes had contaminated a few of Woodward's samples. The various comments in the May edition of *Science* all pointed in the same direction. However meticulous Woodward's experiments, they had not been immune to the contamination problems to which ancient DNA analyses were particularly susceptible. The DNA from dinosaur bones took as swift a plunge in credibility as had the *Magnolia* leaves. As with those fascinating Clarkia beds, the plunge left a lot of questions unanswered. Scott Woodward's bones did seem interesting. What surviving biomolecules were contributing to the waxy texture? What were those surviving cellular structures, and what molecules within them were picking up the microscopy stains?

We now know enough about the chemical fate of bones to show that it is unlikely DNA will remain accessible to our current PCR methods in bones as old as these. The collagen matrix of the bone would long since have turned to soluble gelatine, leached away and opened up the bone's interior to water and air. Only at the very lowest temperatures could bones have escaped this destiny, and bones more than 2 million years old are not going to be found in permafrost. If the waxiness of Scott Woodward's Cretaceous bones indicated some kind of lipid accumulation, then some adjustments to this rule may be necessary. However, there is no evidence of this, especially now that it seems that none of his forty-two samples yielded anything other than contaminant DNA. More of a question mark hangs over the mineral fraction of the bone, the impure crystals of calcium phosphate or apatite. Matthew Collins, who has made a close study of the chemistry of bone, has speculated that the mineral surface itself provides the most durable depository for long-chain molecules such as DNA. If they survive here, as some are convinced they do, then they survive because they are very tightly bound. Here lies the problem. The same tight protection that may prevent chemists and biologists from seeking them out in the normal course of events also prevents analytic scientists from separating and identifying them. In the final chapter, we shall return to some recent attempts to overcome this problem. In the meantime, however, the time frame for ancient DNA analysis stretches back not much further than the Siberian woolly mammoth, 50,000 – perhaps as much as 100,000 – years.

This time span takes us back as far as the final stages of the human evolutionary drama, and the disappearance of the last hominids other than our own species. It extends beyond the appearance of what prehistorians call 'modern behaviour', vividly expressed in Stone Age rock art. It encompasses a period in which humans changed the genetic configuration of several other species, fuelling their most significant transformation of the ecosystem, through agriculture. Within a time scale that is shorter than the dinosaur hunters had hoped for, but which is long in terms of our own species' history, ancient DNA can reach widely into the archaeological record. In just one lifetime, both archaeology and the life sciences have hurtled forward. Within living memory one discipline was all pots and stones, the other long lists of Latin names. Now they both converged on the code embodied in life's most remarkable molecule, and the information flowing from variations in a sequence of four chemical bases. Like the range of text that can flow from a mere 26 letters in the alphabet, that information seemed boundless.

3

our curious cousins

the far reach of ancient DNA

Tomas Lindahl had suggested a figure of between 50,000 and 100,000 years as his prediction of how far back accessible and amplifiable DNA might persist within a relatively cold bone. So where does that take us in terms of the human past? It goes beyond those most evocative expressions of the human mind encountered in cave art: these images appeared in different parts of the world less than 40,000 years ago. It also reaches back past the remarkable diversification of toolkits that occurred around the same time. It precedes any evidence of musical instruments and the elaborate rites of passage indicated by the treatment of the dead. Many of the key features that in our minds separate the 'cultural' humans from their more 'natural' predecessors appear in the archaeological record less than 40,000 years ago. The oldest DNA reaches back beyond that, to reveal an earlier episode in the human past.

Although the attributes of human culture listed above are not encountered in these earlier deposits, bones and a less diverse range of stone tools can be found. Recognizably human skeletal remains can be found in various parts of the Old World. They have no particular physical traits that cause us to place them outside the range of modern human forms. As far as we can tell, they are members of our own species. At either end of the Old World, in Europe and in east Asia, something else crops up. Today there is no question about what is human and what is not. We are a single species, with no barriers to fertility within our own species, and an absolute barrier to fertility between all other primates and ourselves. The position was less clear

50,000 years ago. In south-east Asia, skulls can be found similar to our own, except with a flattened skull and strong brow-ridge that no living human shares. In west Asia and Europe, a prominent brow-ridge appears in another group of specimens. Such traits survive in south-east Asia until at least 27,000 years ago, as new scientific dating methods have recently shown, at a site along the Solo River at Ngandong, Java. The other bearers of strong brow-ridges were still alive 20,000 years ago further west in southern Spain. Archaeologists and anthropologists have long debated who these curious cousins were, and how different they were from us. What was a meeting with one of these hominid relatives like? Did the experience have the separateness of encountering a chimp or orang-utan, with a similar shape but clearly of another species? Alternatively, was the meeting something more intimate, perhaps with exchange of things or ideas? Might it even have led to one bearing the other's children?

This issue has taxed archaeologists and anthropologists ever since the bones came to light. First it was tackled simply by looking at the bones themselves, then attention moved to the stone tools and the archaeological sites they left behind. It has not been an issue simply about a few long-dead relatives. Underlying the issue of how different we are from our curious cousins is a far more fundamental debate about the evolutionary dynamic of the natural world around us, and the nature of our place within it. To understand that underlying debate, and the place of ancient DNA in resolving it, let us look back to the bones from which those fragmentary molecules came.

once upon a time in the west

The bones in question were unearthed in 1856 from the Feldhof Cave, above the Neander valley near Düsseldorf. To the quarrymen who found them they appeared almost human – but not quite. There were a number of significant differences. The thigh-bones were too thick and curved, and most strikingly of all there was a sizeable ridge of bone protruding from the brow. A local schoolteacher was alerted; fortunately he realized they were something quite unusual. Speculation ensued. Perhaps they belonged to a congenital idiot, or a foreigner.

The tough bandy legs were reminiscent of horse-riding Cossacks, the intense brow-ridge brought stress and disease to mind. These were weak explanations for highly distinctive features. What is more, these bones were not the only unusual things coming out of the ground at the time.

As Christians, those quarrymen had very likely learnt that God had created the world at nine o'clock in the morning on the 23rd of October in the year 4004 BC. Yet as the scale of quarrying grew with the progress of nineteenth-century industry, things were appearing from beneath the Earth's surface that were increasingly difficult to contain within this story. Sediments were cut through that surely had taken a great deal longer than 6,000 years to accumulate – but maybe geologists had got their timing wrong. Fossils of weird animals appeared – well, maybe Noah's Flood could account for their absence today. Occasionally, those animals were found at the same level as worked tools – this was getting tricky, but could we be sure those hand-axes were fashioned by humans? As the quarrymen dug deeper, so the biblical counter-arguments became strained. With the Industrial Revolution continuing apace, quarrying on a considerable scale, and the counter-arguments at the point of collapse, the sheer weight of evidence for a deep and complex Earth history was crying out for a different story. Now, in a valley called Neanderthal, the quarrymen had found something that seemed to be only partly human.

It was only three years later that Charles Darwin was to bring the whole creationist story to its knees. In its place, he offered a new, coherent story into which all the new information arising from decades of digging into the Earth – slow sedimentation, fossils, extinction, and species transformation – could reasonably comfortably fit. The story was one of evolution by natural selection. It explained how, over a very long period of time indeed, going back far beyond 4004 BC, species of plants and animals changed in form, as a result of a rather unpleasant process, the struggle for survival. Over the next few years, Darwin and his colleague Thomas Huxley hammered home the implications for humanity. We were not created *ab initio*, but had instead descended from apes. A century and a half later, the shock waves have yet to settle.

The find of Neanderthal man played, and continues to play, a key role in this new world-view. Huxley argued that its strange form was to do, not with idiocy, horse-riding or disease, but with evolutionary change. Here was a 'half-human' who bore direct witness to the mechanisms that, Darwin argued, had underpinned natural history. The bones had only remained a puzzling enigma for a few years before taking their place as part of the foundation of a radically new view of the world and of our place within it. The varnished bones became an important museum specimen.

Once the idea of a half-man midway between animals and ourselves had sunk in, attention turned to the novel concept of linear, directed evolution. The quarrying activity that had exposed so much of the new information was part of a wider economic revolution. Those at the heart of the revolution warmed to the optimistic idea of universal human progress, a progress of which they now occupied a vanguard position. The gradual evolution that Darwin had explained formed a basis for their status, for their premier position in the natural order of things. Deep in the mists of time were undifferentiated species without specialized senses, without sensibility. Slowly but surely, they acquired backbone, structure, a nervous system and brain. Then they stood upright, became self-conscious and acquired dexterity and language. Ascending this path, climbing the evolutionary tree, they could still see the offshoots of these various stages, whose progress had stopped at a lower level: worms, insects, vertebrates, mammals and apes. At the tree's crown were those epitomes of progress, humans. If one looked more closely at the crown, the process of natural selection could be viewed in detail. Some humans were a little below the crown's tips, looking up with awe to the Victorian men of capital and enterprise who embraced the new 'Darwinism' which in turn endorsed their position closest to the sun.

The Neanderthals, as those skeletons closely resembling the original find from Neanderthal came to be called, were well beneath the shade of the crown. As the years progressed, other lowlier beings were added to those shady positions. In the 1890s, a Dutch doctor found bones of what would later be called *Homo erectus* on the island of Java. These were the skulls with the strong brow-ridge that had persisted to a late date in south-east Asia. In 1907, a jawbone was recovered from Mauer, near Heidelberg, from a species that would later be named *Homo heidelbergensis*. Elsewhere in Europe, the number of Neanderthal finds was growing. Indeed, some examples had clearly come to light prior to the Neanderthal Valley example, but simply a little too early for the concept of a 'proto-human' to be considered. However, the Neanderthals have never perched very comfortably on the evolutionary tree directly below us, for a variety of reasons, not least because on average their brains were larger than ours.

At first, these linear stories of evolution and progress had addressed some of the mismatches between the biblical account and the data emerging from quarrying beneath the ground. But, as that database grew, so simple linear evolution came under strain. The diversity found in nature increasingly resisted being ordered in a single straight line. All the elements of Darwin's own mechanism of evolution have survived, and still survive in a robust state, but they needed a 'bush' rather than a 'tree' to turn them into an effective account. Each group of plants and animals led its observers along a similar line of reasoning. There was so much variety in nature, it simply couldn't be constrained within a single axis of variation. There had to be offshoots, side-branches, dead ends all over the place on this spreading evolutionary bush. That was a reasonably painless modification of the story when dealing with insects, starfish, ferns or whatever, but when it came to humans something still jarred.

Even having been absorbed into Darwin's biological world, we have remained in awe of ourselves and our species. It has seemed to us that our powers of perception, our conscious thought and creativity, and our sheer ability to control nature, are attributes of such magnitude that they are a case apart. How could our hominid cousins, so close to us on the evolutionary bush, avoid being drawn on to our path? So much human evolution continued to read like a multi-millennial quest to lift those forelimbs off the ground, to open up that brain cavity in preparation for thought and to stretch that larynx out in anticipation of speech, so that music, art and civilization would one day blossom forth. The subtext of human evolution remained a drive for progress, a collective struggle for betterment of the species. In explaining these fossil bones, both the Neanderthals of Europe and the late *Homo erectus* fossils from China and Java were repeatedly drawn back into the story. This continued to be the case, even as other evolutionary trees were changing from skyward pines to untidy bushes. Franz Weidenreich, writing in the 1940s, and Carlton Coon, writing in the early 1960s, were able to organize the known hominid fossils into one global story in which the watchwords were ascendancy and slow, gradual progress. For these authors, the Neanderthals, Peking Man and Java Man were all part of one common species. The variety between them reflected deep-seated racial differences across the different continents of the world, differences that have endured until the present day.

As they developed their ideas and brought them into print, another continent was yielding up a remarkable series of hominid fossils that could be taken to suggest a quite different story of origins. As Coon was preparing his manuscript, the African hominid record was about to explode, as Louis and Mary Leakey started to find the first of a series of key fossils where the Rift Valley split the Earth open at a gorge called Olduvai. As scientific dating methods came on stream to put these fossils in order, so their dates as well as their form changed the global focus of human evolution. Europe and Asia became outliers to the main action. The earliest dates were to be found in East Africa, as was the greatest diversity of fossil types. An alternative to the multi-regional argument was emerging, which gave primacy to one particular continent, Africa.

enter the molecules

Something else was happening in the 1960s that would eventually point in the direction of Africa. While the Olduvai Gorge was yielding fossils that would change the whole story, a young New Zealand scientist called Allan Wilson was wondering whether these fossils on their own could hold the key. Wilson's laboratory would subsequently hold a central place in the growth of ancient DNA science. Back in the 1960s, he was one of the pioneering figures to shift the focus from whole fossils to molecules. At this stage, the double helix was little more than a decade old. It was too early to explore its fine structure as we are able to do today. At that point, Wilson examined a better understood group of molecules, the proteins. He was growing sceptical about the way primate anthropologists were trying to relate different fossils according to differences in skeletal form. The small sample sizes worried him, as did the vagueness about the links with genetics, and with the difference between sexes. Neither did he like the jargon physical anthropologists used, which struck him as a bit of a smoke-screen. Molecular science had moved forward enough to take what seemed to him a more transparent and testable approach. Together with Vincent Sarich, at the time a graduate student in anthropology, he attempted to uncover the family tree of humans and other primates by comparing the structures of serum albumins, the predominant proteins in blood plasma. What they did was develop some antiserums to a human serum albumin, and then test the extent to which they cross-reacted with the serum albumins of a whole range of other species. None cross-reacted as if it was identical to humans but a fairly strong cross-reaction was achieved with chimpanzees, regarded by primatologists as our closest living relative. Other apes, including the gorilla, orang-utan, gibbon and siamang, also gave moderate cross-reactions. The strength of the cross-reaction progressively weakened as the tests moved to Old World monkeys, New World monkeys, more distant primates, and non-primates. The molecular variation among the albumin proteins seemed to make evolutionary sense, and open up a new route to our phylogeny. So far, so good – but things quickly became more contentious with another property of the molecules. They could be used to generate a measure of time.

The pace of evolution can vary. Some disease organisms can evolve into new forms faster than we can develop remedies against them, while a marine invertebrate called a brachiopod has retained much the same form for hundreds of millions of years. There is no in-built evolutionary rate for organisms as a whole. Molecules are different. Their change is more directly governed by rates of chemical reaction. Those rates may vary, for example according to the temperature and the presence or absence of catalysts, but they remain rates that are broadly determinate.

Let us take the example of the serum albumins studies by Sarich and Wilson. The strength of the cross-reaction between human anti-serum and chimp serum provided a measure of the sequence similarity of two proteins that performed precisely the same function, but in different, albeit related, species. Those differences are not the outcome of differential natural selection, as nothing functional has changed. Instead they are to do with accumulating chemical aberrations, or mutations, along bits of the molecule away from the main functional areas, away from the powerful forces of selection that would otherwise erase them. These mutations simply accumulate like dust on an unswept surface, away from the main ecological action.

If chimps and humans do have a common ancestor, then there must have been a particular point in time when the two lines of descendants began to diverge. From that point onwards, the random mutations along the chimp line will accumulate independently from the random mutations in the human line. So long as there is some steady probability of mutations occurring through time, the number of sequence differences will be a measure of how long ago that divergence happened. This is the logic that Sarich and Wilson employed. By considering a variety of species, they demonstrated that the levels of difference between albumin proteins seemed well ordered, and came up with series of time estimates for different episodes in the human evolutionary story. Drawing on all the measures of albumin dissimilarity among the primates, they concluded that the common ancestor of all primates was around six times as ancient as the common ancestor of humans and our closest living relatives, the chimps. The oldest fossil Old World monkeys go back 15–20 million years, and combining fossil and immunological arguments the authors favoured a primate ancestor at 30 million years. This implied that we split from the chimp and gorilla lines as recently as 5 million years ago.

Eyebrows were raised across the hominid fossil fraternity who had been getting increasingly confident about their own scientific dating methods and their broadening range of hominid fossils from secure, dated contexts. One small assemblage of fossils came from the Siwalik Hills of northern India. It was labelled *Ramapithecus*, and dated to 14 million years before the present. There seemed to be a clear contradiction between this tangible, datable fossil and the rather novel argument from a quite separate discipline, about albumin proteins, that seemed to float free of the archaeological record. The scholars who knew about the fossils did not feel ready to collapse their global picture of human evolution to a particular part of East Africa. Neither were they persuaded by a time scale that had shrunk to a third of the age of its oldest fossil. They were not prepared to ditch the received wisdom of human evolution as a story of long, slow progress. Sarich and Wilson initially came in for a lot of criticism.

They only had to wait a few years, and for the recovery of new specimens, for the fossil evidence itself to remove its own barriers to the shorter time scale. In 1976, David Pilbeam found a better *Ramapithecus* fossil near to the original find, and detailed enough to allow the species to be gracefully 'retired' to the orang-utan line. In the subsequent years, the time estimate of 5 million years for a common ancestor with chimps has broadly survived, and given great strength to a powerful tool of molecular evolutionary science, the so-called 'molecular clock'.

molecular clocks

Archaeologists and earth scientists looking back into the past are accustomed to working with a variety of natural clocks. The basic requirements of such clocks are that they proceed at a knowable pace, as do very many natural processes, and that their pace does not get deflected by what is happening around them. The most important of these are the so-called radiometric clocks. These use radioactive breakdown in one form or another, suitably triggered by some archaeological or geological event. Much of the dating of early hominids, for example, has made use of the slow breakdown of one form of potassium to argon gas, a valuable means of dating volcanic deposits in Africa's Rift Valley. Similarly, the prehistory of the last 20,000 years or so is now reconstructed around carbon dates, derived from the slightly faster breakdown of a certain form of carbon to nitrogen gas. Archaeologists also make use of a series of chemical processes to supplement these radiometric timekeepers. An example of these is the measurement of a layer of hydrated silica that gradually deepens on the surface of stone tools after exposure. In some respects, molecular clocks are similar, involving a triggered, time-dependent chemical reaction. In other respects, they are rather different. First, the chemistry behind mutations and altered DNA and protein sequences is very complicated. The same could be said of radiometric dating, but here the complexities are rather better charted and understood. We would not expect molecular clocks ever to come anywhere near to rivalling the precision possible with radiometric clocks. Second, molecular clocks, as part of reproducing organisms, are subject to natural selection. Chance mutations may progressively distance the molecular sequences of two relatives, but we can rarely be sure that natural selection will not then steer them back again towards an ecologically favoured format. The basic processes of evolution can thus distort and confuse the time scale used to measure its pace.

This is invariably true of any biomolecule with a significant influence on the form the whole organism takes. In whole organisms, redundant structure is uncommon. Everything seems finely tuned by the brutal rigours of natural selection. There are no spare limbs to be found and hardly any dispensable organs. This forced economy of whole-organism design has always limited the use of bodily form as an evolutionary timepiece. If selection pressure is strong enough to mould one lineage in a certain way, then many lineages may be so moulded. A shared physical feature may either result from common ancestry, or from convergence in a tough, competitive environment. It is not always easy to separate the two.

Proteins are not totally immersed in this environmental drama. True, they each have their functional regions somewhere along their length, conducting precise, crucial biochemical tasks. However, there does seem to be considerable arbitrary baggage within these long-chain molecules. It is not always easy to draw a line between functional and non-functional sequences, between the stretches faced with natural selection and those undergoing 'neutral', that is random, evolution. Allan Wilson was acutely aware that this could apply to molecules as much as it did to whole organisms. Later in his career he published some elegant examples of convergent evolution among proteins. To approach a more ideal molecular clock, the molecules had to distance themselves from the environment and from natural selection, and that is where DNA comes in.

the surprising world of the DNA molecule

As methods of sequencing DNA molecules and reading their genetic code came on stream in the 1970s, the tacit assumption had been that the existence of DNA was subservient to that of the whole organism and its evolutionary battles. One of the surprises was how little of the stuff seemed to be engaged in those battles. It transpired that 'coding genes' – the sequences of DNA actually used to build proteins and thus engage in the outside world – make up as little as 10 per cent of the DNA within the cell's nucleus. The other 90 per cent reproduce without any connection to the trials and tribulations to which the whole organism is exposed. These emerging non-coding regions were ideal for charting evolutionary patterns independently from natural selection. This is not to say that previous evolutionary markers were displaced by this, but there was a progressive distancing that could be achieved from natural selection and the confusions of convergent evolution. This moved first from whole organisms to proteins, and then on to non-coding DNA.

These non-coding regions are the best source we have for independently tracking lineage and generating a molecular clock. We can predict that on average a particular stretch of DNA is likely to accumulate a new change every few hundred generations, or few thousand years. This will not proceed like clockwork. As a random process, it is not entirely predictable, but over the long term this randomness is absorbed into a relatively uniform rate. If we focus on that particular stretch, then two closely related individuals will have almost identical sequences, and increasingly distant relatives will have increasingly disparate sequences. The sequences thus form the basis for the construction of phylogenetic trees, but this is not an entirely self-contained process. At some stage, these molecular projections of evolution had to be anchored on some real dates, bringing them back to the archaeological and geological records.

The way the molecular clock is normally calibrated is either by reference to the fossil record, and some well-dated assumed common ancestor, or by reference to some well-dated separation of land masses, marking when close relatives either side of the barrier began to diverge. One of the important features that such reference points reveal is the small number of mutations involved in measuring archaeological time scales. This is a key feature of the molecular clock. It is not so much wrong, as imprecise in the extreme. Many of the issues about chronology discussed in this book push the clock well beyond the limits of its resolution. It thus stands as a first rough estimate, often leading to a need for much sharper archaeological dates.

By the early 1980s, an important step forward in the endeavour of DNA sequencing would open the way to exploit fully the possibilities of applying a molecular clock to human ancestry. For the first time, one whole section of the human genetic code had been fully sequenced, with every single base-pair along the double helix charted.

mitochondrial DNA

Cracking the genetic code presented a breathtaking task to the genetic sciences. It was as if they had unearthed the keys to the world's most prolific libraries, all the volumes of which had yet to be read. But where should they start? The DNA within the nucleus of the human cell is enormous, made up of 2–3 billion base-pairs distributed among twenty-three pairs of chromosomes, and coding for between 30,000–40,000 genes. Imagine the number of letters in all the words in all the volumes in a small library. That is the order of magnitude of the total sequence. Its decipherment has been the goal of the international Human Genome Project. In the early days of DNA research, there was a less daunting task available for study.

Situated in the cell fluid outside the nucleus are its powerhouses, the mitochondria. The number varies according to how much work the cell does. Not surprisingly, a muscle cell has quite a large number. These small powerhouses also have a small amount of DNA within them which, like the more prolific quantities in the cell's nucleus, contains the blueprint for certain of the cell's molecular 'machine tools'. Unlike the nuclear genome, which is shuffled with every sexual reproduction, so that offspring receive genetic information from both parents, mitochondrial DNA is normally inherited from the mother alone. Its sequence is not reshuffled during reproduction, and changes only as a result of occasional mutations and any selection pressure that may act upon them.

Not only does it have this simpler system of inheritance, without recombination, but it is also much smaller than the nuclear genome – 100,000 times smaller in the case of humans. If the nuclear genome is a small library, the mitochondrial genome is a large pamphlet. It thus posed a more achievable target to the first generation of DNA sequencers. By 1981, the entire human mitochondrial genome had indeed been sequenced. Three successive pages of the April edition of *Nature* for that year carried a densely packed sequence of the letters A, C, G and T, corresponding to the order of 16,569 base-pairs along one of the two strands of DNA. The other strand carried the complementary sequence of bases with which the first strand's sequence was coupled.

Printed above some parts of the sequence was another series of letters, this time corresponding to the proteins built from this blueprint. Yet further parts of the sequence were boxed in, and related to the genetic functions of the sequence. It was an extraordinary achievement and generated a genetic map that would serve as the basis for a whole series of research initiatives, including the analysis of ancient DNA. Subsequently, the mitochondria of mouse, cow and rat were also sequenced, and they proved to be very similar. In other words, this new genetic map was of value, not just in the case of humans, but also in exploring other animal species.

The shape of the human mitochondrial genome is quite different from that of the nuclear chromosome. While the latter is a thread-like object, made up of linear DNA molecules tightly and intricately bundled together with proteins, the mitochondrial DNA is a rather simpler, circular molecule, the two interconnected strands continuing round without a break. Comparisons with mouse, cow and rat also show it to be evolving ten times faster than the nuclear genome, a feature that lends itself well to exploring small evolutionary differences. The numbering system that Anderson and his colleagues used to locate any partial sequence on the circle therefore had an arbitrary origin, but situated sequences in relation to each other. Take the example, considered in the previous chapter, of Scott Woodward's search for dinosaur DNA. He had targeted the region between base positions 15,603 and 15,777 on Anderson's map, which lies in the middle of a gene coding for a protein called cytochrome b, one of the cell's energy-processing molecules. Unlike the nuclear DNA, in which coding genes are rather dwarfed by the amount of non-coding DNA, human mitochondrial DNA is tightly packed with genes and the non-coding regions are in the minority. A number of these coding sequences build proteins like the one above which are involved in the cell's energy management, a suitable role for the cell's own powerhouses. Other regions of this circular sequence coded for various kinds of RNA structures, the molecular tools used in protein building. Different parts of these RNA structures can mutate and vary without damaging their function, and have been examined to explore certain relationships between taxa. However, the part of the genome that has attracted most attention by ancient DNA analysts is a sequence of just over 1,000 base-pairs that appears to have no coding function at all.

Reading around Anderson's linear map of the circular sequence, this non-coding region falls between positions 16,026 and 00,577 (the sequence continues from position 16,569 to 00,001). Although this region has no coding function, it has a number of interesting features. First, a number of the trigger points that control transcription, that is the reading off of the blueprint to make proteins, occur along this sequence. For this reason, it is sometimes referred to as the control region. Second, there is a rather unusual structure called a displacement loop, or D-loop, within it. This loop is a sequence of about 680 base-pairs where the double strand opens up to form a kind of eye along the length of the circle. One strand has been displaced from the main circle as a result of the other strand pairing up with its own partner strand of RNA, in this way displacing the loop of single-strand DNA.

The control region and the D-loop within it, in all making up almost 7 per cent of the entire mitochondrial sequence, has been a key target for those exploring close evolutionary relationships by modern and ancient DNA alike. Two or three sections within the control region contain the most variable sequences in the entire genome. In some parts of the genome, the greatest mutation tolerated is an occasional base replacement in a position that is not going to impair function. In these highly variable or 'hypervariable' segments of the mitochondrial genome, not only do the bases change, but even the length of the sequence changes. This is shown dramatically in what is known as the first hypervariable segment of the mitochondrial control region. This stretch lies at one side of the D-loop. So fast is it evolving that not only its sequence of bases, but also its actual length varies considerably between species. In humans, it is a little below 400 base-pairs in length (positions 16,023 to 16,400), in rats just below 300, and in cows below 200 base-pairs. This is a quite remarkable level of diversity for what are quite closely related species. Even within a single, relatively young species like our own, this segment displays marked variation. Human sequences will differ among themselves by an average of eight substitutions along the segment. This first hypervariable segment presented itself as an ideal place to look for relationships between closely related organisms, such as within a single genus or species. In time it would crop up again and again in ancient DNA analyses, and was soon to have an impact on the human story in relation to an ancient woman who came to be dubbed 'the mother of us all'.

mitochondrial eve

With Anderson's mitochondrial sequence to hand, Allan Wilson was now able to apply the same logic to DNA that he had earlier applied to albumin proteins in the 1960s, but with a sharper taxonomic precision. Variation in the human sequence would reveal something about human origins. Two of his graduate students, Rebecca Cann and Mark Stoneking, set to the task. Typing DNA in the early 1980s still required large quantities of tissue. The potential of PCR had not yet been realized and so there was no question of typing DNA from a hair or a small drop of blood as can be done today. Cann and Stoneking worked instead with whole human placentas. About two-thirds of these came from American hospitals, and others came from Aboriginal populations in Australia and New Guinea. The total sample of 147 specimens included 2 Africans and 18 African Americans, 34 Asians, 46 Caucasians, 21 Australians and 26 New Guineans. Their mitochondrial DNA was purified and analysed and the different types compared. These comparisons were made in relation to the different regions of the mitochondrial genome, within the D-loop, the protein-coding regions, the RNA-coding regions and elsewhere. Little more than a decade after Wilson's lab had changed the human story with one set of genetic patterns, his students would now change it with another. Once again, one of the key features in the new story was a matter of time scale, and a pace of evolution that would further distance us from our curious cousins.

The central findings were that the variation was very little, in comparison to other primates, and most of the variation could be seen within the African Americans. Taken together, the results implied that all living humans could trace their mitochondrial lines back to a common female ancestor around 200,000 years ago. Most of the existing variation within the sequence was found among Africans. That suggested that the single human lineage also found its root in Africa. The mother at the source of that lineage captured the popular imagination as 'Mitochondrial Eve'.

Taking these two points separately, the small amount of variation resonated with the Wilson lab's earlier finding. Just as the albumins had shown that the hominids were more recent than the traditional story had held, so the mitochondrial DNA indicated a recent time scale for our own species within the hominids. There had not been time for a lot of variation to accrue. We saw earlier how these abbreviated time scales were difficult for those who wanted to model all the Asian fossils within a single gradual story of inclusive and progressive evolution. The second feature of Cann's and Stoneking's results accentuated that difficulty. The DNA results placed that recent ancestral 'mother of us all' in Africa, marginalizing a series of fossil records scattered around Europe and Asia which others remained keen to retain within the central human lineage. Around the world hominids could be found who significantly pre-dated the short time scale proposed by Wilson's group. Moreover, the dates were safe. Accepting the out-of-Africa story certainly meant the end of the great collective drive towards human progress. It would mean hominid species more often than not went into extinction. This was not so surprising given that the fossil record indicated the same was true of most species, but was nevertheless a profound challenge to the still-persisting view that we humans are different.

The multi-regionalists, as those arguing for the inclusive approach are known, responded to the Mitochondrial Eve hypothesis in a predictable manner. The criticisms came from a new generation of physical anthropologists. Foremost among these were Allen Thorne and Milford Wolpoff. Both authors had been struck by physical variations between the fossil skulls in the eastern part of the Old World range. It seemed to them that a group of the earliest east Asian hominid skulls displayed features that could still be seen in human populations alive in that part of the world today. Likewise, they linked certain attributes of south-east Asian fossil skulls with recent and modern Australian forms. Their multi-regional theory was not a direct replay of the earlier stories of Coon and Weidenreich. They had by now accepted both the shorter 5-million-year time scale for the hominid line, and the original source in Africa. But they were still looking for a model that drew all the Old World fossils into a single collective story of forward evolution. Attaining that required a time period five times what Rebecca Cann's results permitted. They vigorously sought out the weak points in the arguments.

The clearest weakness was in the sampling, particularly of 'Africans'. With two exceptions, the placentas were taken from African Americans. Let us take just one hypothetical mother and conjecture that seven out of eight of her great-grandparents were of African descent, but that her mother's mother's mother was European. Because of the way that nuclear chromosomes work, her physical features would be unlikely to display an undue influence from that 12 per cent of her DNA. The 88 per cent of African descent is what we would discern. However, her mitochondrial DNA will be 100 per cent European, and that would obviously skew the results of Cann's and Stoneking's analysis. In this particular instance, the analysis was probably on reasonably safe ground. This is because in America, so far as we know, the great majority of marriages between those of European and African descent have tended to be between European men and African-American women. None the less, the principle holds. We could repeat this logic for a whole range of American ancestries, just to make the point that mDNA lineages in a multi-ethnic society may not be directly transposable to the rest of the world.

This in turn related to a much wider perceived weakness in the argument surrounding mitochondrial lineages. Such lineages are rather like the female equivalent of surnames. Each is passed between generations undiluted and from only one parent, thus reflecting a decreasing fraction of the genetic contribution of an increasingly distant ancestor. Let us take the analogy further. Thumbing through page after page of my own extremely common Welsh surname in an English telephone directory, one might have cause to develop an 'Out-of-Wales' hypothesis for the peopling of England. I know the surnames of eleven of my sixteen great-great-grandparents. They are all different. Few are Welsh, and just as many are of Italian origin, but Londoners outnumber the two put together. The phylogenies, first of the surname Jones, second of my mitochondrial haplotype, and third of myself are three distinct phylogenies, and have no need to match up. Thorne and Wolpoff were of the view that mitochondrial evidence served an interesting function of generating hypotheses, but those hypotheses could only be ratified or refuted by reference to the fossil and archaeological record itself. In their view, those records refuted it. They saw no evidence of a rapid replacement of one hominid group by our species, and no evidence of the uniformity one might expect to result from such a replacement. The hypothesis, however interesting, did not fit with the data and there was therefore no problem in rejecting it.

The second weakness arose from the use of the molecular clock. The sceptics were still uncertain about Wilson's earlier use of sequence variation to derive a figure of 5 million years. Now the molecular clock was brought into play for an even shorter time span. Opinion still varies as to how far it can be pushed to perform in the very recent evolutionary time scales of archaeology. Thorne and Wolpoff doubted it could be brought forward more than half a million years.

If that was not enough, the use of the computed tree was also an interpretative minefield. Computers are powerful things; it's not so difficult to get a computer to take a batch of data and build a tree out of it, based on some variable related to similarity. It is not even so difficult to get it to build a range of different trees, by fine-tuning what is meant by 'similarity', and adjusting other variables and assumptions. The multi-regionalists quite predictably went for that Achilles' heel with vigour, and repeatedly found a target to hit. It was a baptism of fire for Rebecca Cann, who experienced the furore that could arise from challenging established views about the human past. The comment ranged widely between enthusiasm and ridicule, but grant applications became difficult. But for the encouragement of her supervisor, Allan Wilson, Cann was ready to quit. To him it was a familiar scenario. His compressed time scale for primate evolution had caused a similar rumpus twenty years earlier, but had stood the test of time. The Mitochondrial Eve argument, and all its implications about the wider pattern of human beginnings, had some way to go.

While it may be tempting to portray the conflict as between the evidence of fossils and that of molecules, it would be a false portrayal. Many of those studying the fossils have concluded that the details of their variability fit best within the recent 'Out-of-Africa' model. They also had the contentious molecular results at their disposal. Both sides persisted with the case, until one more piece of information was extracted from the fossil bones. The growing collection of almost human skulls from around the Old World had been scrutinized for ever finer details of form and structure. What was needed was for one of them, be it one of the south-east Asian *Homo erectus* fossils or a European Neanderthal, to yield up a clue about their own genetics. As fate would have it, it was the very first of these fossils to be recognized that would oblige.

neanderthal genetics

Allan Wilson, whose research group had been central to this story of exploration into the human molecular past, tragically died before its culmination. His colleague, Svante Pääbo, had recently moved to the Munich Institute of Zoology to take up a professorship. A few hours' drive to the north-west of Pääbo's new home was the cave above the River Düssel in which the remains of the first recognized Neanderthal had been unearthed. They were now kept in the Rheinisches Landesmuseum at Bonn, where Ralf Schmitz and his colleagues were engaged in a new study of the Neanderthal specimen. Pääbo immediately opened negotiations to sample the specimen for ancient DNA. He had to be patient. The museum curators were very rightly cautious about the request not simply to study the bones, but to sample one of them destructively for a technique that was still very new to the scientific world. The age of these bones was still unclear. They were somewhere between 30,000 and 100,000 years old, which was clearly pushing DNA recovery to its limits. A likely outcome of cutting out a sizeable chunk from this precious specimen was a nil result. Balanced against this was the potential, for the very first time, to explore the genetics of a hominid other than ourselves. After several meetings, and the passage of a few years, the curators agreed that the work was sufficiently important to give it the green light. The meticulous investigation of Pääbo's team demonstrates how far biomolecular archaeology had come by the mid 1990s. In place of a speculative dive into poorly understood ancient tissue, theirs was a careful analysis of biomolecular preservation of the bone as a whole, gradually homing in on the ancient DNA, followed by cross-checks and a critical phylogenetic consideration of the results.

Pääbo's team sawed into one of the long bones, removing a segment upon which the analyses could begin. They first looked at the state of the bone's proteins, and of the proteins' molecular building blocks, the amino acids. This was in order to get some sense of the general molecular survival within the fossil, before embarking on the target molecule, DNA. The levels of amino acid in the ancient bone were between one-quarter and three-quarters of the equivalent levels in a fresh bone. This was a very encouraging result, and they went on to examine the state of the amino acids in order to get a further idea of the conditions of molecular preservation. We saw in the last chapter how Hendrik Poinar had worked out how to use amino acid racemization as a measure of how dry the molecules had remained. They tried it now on the Neanderthal specimen. The level of racemization was low, indicating that the heart of the bone had remained almost completely dry.

At first, this result may surprise. Why should a bone, for thousands of years in a fairly ordinary sediment, have stayed dry? Moreover, there are now a number of successful ancient DNA amplifications from bones recovered from waterlogged peats, which certainly were never dry. The truth is that the bone surface can remain moist, but islands of solid tissue within can remain dry. What matters for molecular preservation is not necessarily the general state of the tissue, but of the best conserved components of it. As molecular techniques have become fine-tuned to deal with smaller and smaller samples, so those conserved islands of molecular survival could also get smaller and smaller, yet still allow detection.

The solidity of an ancient bone has a lot to do with the temperatures to which it has been exposed. At too high a temperature, the collagen, one of the molecules endowing the bone with strength, would turn to gelatine and dissolve, leaving a minutely porous structure open to breakdown. The Feldhof cave was well to the north of the Neanderthal range, close to the limits of glaciation during the cold spells of the Quaternary Epoch, and clearly a good place to start looking for molecules. Hendrik Poinar's amino acid assays looked encouraging; the interior of the bone had remained dry. The team set about their hunt for ancient DNA.

In a specimen so old, the DNA was bound to be heavily fragmented. The team needed to target a short length, but one with the right level of variability. The first hypervariable segment of the mitochondrial control region was clearly a good place to look, for the reasons discussed above. They chose a 105-base-pair segment of the first hypervariable segment at one end of the D-loop, and set to work designing primers. After a series of PCR cycles they separated out the amplified DNA and got a positive reading at the level corresponding to 105 base-pairs. A century and a half after these curious bones with their enlarged brow-ridges and sturdy limbs had first taken the world by surprise, those same bones were yielding up a short fragment of their genetic blueprint.

After many checks and balances, it was clear the researchers had succeeded in amplifying a DNA strand that was similar to the equivalent strand in modern humans, but too different to be explained as contamination. To be doubly sure, Pääbo sent a sample of bone to his old colleague from the Berkeley days, Mark Stoneking, now at Penn State, to check the result in a different lab on the other side of the world from Munich. Stoneking had some difficulty with the full length that Pääbo's lab had sequenced, but, targeting a shorter sequence within it, managed to corroborate Pääbo's result. In this shorter 30-base-pair sequence, the labs on either side of the Atlantic displayed the same four deviations from the Anderson reference sequence. They were on to something.

Reassured that this really was Neanderthal DNA, and not some contaminant or artefact, Pääbo's team pressed on to a truly ambitious target. They planned to map the entire length of hypervariable segment 1 within the mitochondrial control region. What this entailed was building up, bit by bit, sequences of around 100 base-pairs long, to develop a rather longer master, in this case 379 base-pairs long. This was, in essence, to take the logic of *Jurassic Park* as far as it is technologically feasible to go. In the novel, this same patchwork procedure was employed on DNA from deposits millions of years old, to build an extinct composite sequence, billions of base-pairs long. Pääbo's lab had, more modestly, taken DNA from a fossil tens of thousands of years old, to build an extinct composite sequence hundreds of base-pairs long. The other difference is that Pääbo's team succeeded in the real world.

This hypervariable segment lives up to its name. As we saw above, even its overall length can vary dramatically between different mammal taxa. Within a single, relatively young species such as humans, sequence differences vary from the Anderson sequence by eight substitutions on average. If we compare ourselves with our closest living relatives, the chimpanzees, the average number of differences is 55. The average deviation of the replicate Neanderthal sequences was estimated as 26. This measure of difference was three times greater than the internal measure for our species, and half the difference between chimps and ourselves. Furthermore, it was no closer to modern European sequences than it was to Asian or African sequences. In modern human terms, the Neanderthals were out on a limb.

It was a dramatic finding that seemed to clinch the Out-of-Africa argument. Of all the living humans whose mitochondria had been sequenced, from throughout the world, none of them had anything like a Neanderthal anywhere in their maternal lineage. What Pääbo's team had uncovered was instead a distant relative of modern Asians, Africans and Europeans. It was too far from any of them to be thought of as a close cousin, and with a distinct mitochondrial fingerprint that had never found its way into any modern human lineage. There was no sign of gene flow at all.

These data also provided a basis for a new molecular clock to determine relative evolutionary time scales for chimps, humans and

Neanderthals. Allan Wilson's early use of a protein clock had suggested a common human–chimp ancestor around 5 million years ago, and this figure had not much changed. The most recent estimates placed it between 4 and 5 million years ago. Using that as a baseline, modern human ancestry would come out at 120,000–150,000 years ago, a figure consistent with that of Mitochondrial Eve, and the human–Neanderthal common ancestor lived 550,000–690,000 years ago. The first anatomically modern human that walked had had no relations with the Neanderthal line for half a million years.

The findings were published in the prestigious journal *Cell*, and a news conference was held in London to launch the spectacular findings. The British press hailed the 'coming of age' of ancient DNA research. Chris Stringer compared the scientific importance to landing *Pathfinder* on Mars. Others, while admiring the research, were more circumspect, pointing out that of course this was only one specimen. At that stage it was argued that no ancient skulls of 'modern' form had been tested. Perhaps they too had a very different genotype. An interesting suggestion was put forward that the genetics of early modern humans and Neanderthals should be assessed in Israel, where they co-existed for a considerable time. An early modern form skull has since been assessed by Bryan Sykes in Oxford. It was from a 12,000-year-old deposit from Cheddar Gorge in south-west England. The amplified sequence carries only two deviations from the Anderson sequence, and thus tends to support the Pääbo model. Alas, the Israeli bones are unlikely to be suitable. At Oxford, Alan Cooper conducted a range of essays on Neanderthal and other bones from sites in various temperature zones. He succeeded in amplifying DNA from horse and mammoth bones of similar antiquity, but from Alaska and Siberia. Although the same methods were used, similarly ancient bone samples, including Neanderthal samples, from France, Croatia and Spain, produced a null result.

The original Neanderthal specimen, however, has continued to yield genetic information from its ancient DNA. Two years after Matthias Krings and the Pääbo team published their findings from the first hypervariable segment of the control region, a further analysis appeared in print, this time of its second hypervariable segment. Once again they recorded a larger number of mutations than would be anticipated within our own species. Their earlier findings were further endorsed by a second section of the mitochondrial genome. More recently still, two other Neanderthal bodies, one from the northern Caucasus, the other from Croatia, have yielded up parts of their DNA sequence.

Now these secrets have been so meticulously tapped, we have some insight into how different Neanderthals were in genetic terms, and what that means in terms of the fate of hominid species. While they were so very different, they by no means occupied different worlds. Their common world is something of which we can catch an occasional glimpse.

a tale of two species

It is almost impossible for us to conceive of the vast expanse of time that has elapsed since Palaeolithic hunters last roamed across a windswept frosty plain in pursuit of woolly mammoths, those long epochs during which farming, trade, urban society and the modern age all came into being. Yet for a period of time that was perhaps twice as long as this, two independent species of human coexisted in Europe. Their faces, the shape of their heads and bodies differed far more than humans today differ. They certainly would have seen each other, occupying the same regions and in some cases roaming across the same valleys and places of refuge. Both gathered around hearths to prepare and cook meat, using similar tools of wood and stone in its capture and preparation. Both adorned their bodies, the modern humans at least had language, and some have argued that the Neanderthals played music. Both cared for their elderly, sick and infirm, and buried their dead. The two groups had far more in common than humans share with any living primate, and both had large brains, the Neanderthals slightly larger. There is growing evidence that they watched one another and took ideas about toolkits, adornment, perhaps music and ritual. But they were different branches of the evolutionary bush with separate fates.

It seems that for thousands of years these two distinct species of humans coexisted in Europe. We have no idea how cautious or hostile the two species were, but we now have evidence that they did not interbreed. Ideas passed between them, perhaps even artefacts. In itself, this does not imply a close bond. Today Coca-Cola, transistor radios and nuclear weapons can pass between modern human societies that are far from being on speaking terms with each other. Concepts and things could have passed between the big-brained Neanderthals and our own ancestors in a variety of ways, without them getting into bed with each other. Some Neanderthals did not adopt any traits from their modern human counterparts. Interestingly, these more conservative Neanderthals persisted for much longer than their more malleable relatives. They were still to be found in southern Spain 20,000 years ago, that is 10,000 years after the disappearance of the species from most parts of Europe, and 20,000 years after modern humans first moved into Europe.

In Neanderthals, we see the typical fate of hominids, indeed the typical fate of any species. A genetic combination persists for a certain length of time, but not indefinitely. The Neanderthal had persisted for a few hundred thousand years. Some hominid species may persist for more than a million years, but not forever, and the same will probably be true of us. We are not part of an ongoing and unique surge towards human progress; we are finite components of a natural world that is changing and ephemeral.

For the multi-regionalists, eager to cling to a more idealized view of human progress, the relationship is directly parallel to that between different ethnic groups today. Most reproduction is within the ethnic groups, but with a significant gene flow between them, creating a common species identity and trajectory. The Out-of-Africa model offers something rather harsher, and rather more resonant with the classic Darwinian view of nature red in tooth and claw. There was nothing sacred about human evolution, about the generation of self-consciousness, sympathy and creativity. The hominid species with the biggest brain lost out in the struggle for survival, finally dwindling and disappearing on the south-west perimeter of Europe as its more expansive cousin proved fitter for survival, at least for a while.

Milford Wolpoff has yet to be convinced. He has praised the ancient DNA work on technical grounds, but wonders whether the conclusions are hastily drawn. He argues that just because a gene has disappeared, or a particular DNA sequence is no longer found, that does not necessarily amount to non-ancestral status. Chris Stringer naturally feels differently: 'Of course this is only one specimen, but it fits so well with the view of one side of the argument about Neanderthals – that they are very distinct, that they are not our ancestors – that I think it goes a very long way toward resolving the Neanderthal problem' (News Conference, Natural History Museum, 10 July 1997).

This evidence has captured our own species in its youth, in a lifetime that may only have run a tenth of its course. Just as the archaeological record shows the Neanderthal population both expanding and shrinking, in our case it displays its expansion, out of Africa and across Asia and Europe. The molecules have played a central role in sharpening up that story, but as we move into the most recent 10,000–20,000 years, they play a more vital part. This is when molecular persistence becomes significant for a wide range of species and tissues. It is also in this last brief episode of human evolution, after Neanderthals have left the stage, that the archaeological record of past societies displays its greatest diversity by far.

4

final traces of life

burial

We would be wrong to take the survival of ancient bones, or indeed any organic remains, for granted. Vestiges from the distant past are very variable indeed. Sometimes the search for archaeological remains takes us several metres below the surface. At other times, recognizable traces from some ancient epoch are found intact immediately below the modern topsoil. Rich patterns of ancient, unsuspected structures may be encountered with the removal of as little as a few centimetres of sediment. The fact that bio-archaeologists have anything at all to look at owes a great deal to the simple process of burial.

Had those Neanderthal bones been left exposed on the surface of a fertile soil they would have vanished within weeks, or even days, the last bones stripped of meat and carried away by dogs and birds. Beneath the ground, and, in particular, buried within the sediments of the Feldhof Cave, the story is different. An army of insects, bacteria and fungi can dispatch the soft tissues without much difficulty. With the help of an acid medium around the body, they can also dispatch the bones.

The micro-organisms of an alkaline soil can make much less impact upon sizeable dense organic objects like teeth and bones, and burial is sufficient to keep them intact. Burial also confers protection against the greatest ravages of frost or heat, or fluctuations between the two. There are one or two other organic tissues that burial can protect from destruction on its own. Charred plant remains and the shells of snails and other invertebrates are among these, though their apparent protection is often partial. Their seemingly intact appearance may be an illusion, and the molecules within may be in disarray. None the less, for many excavations, these are the core of the bio-archaeological record: teeth and bones, invertebrate shells and charred plant remains. On sites where the soil is acid, even the bones and shells will have gone. Such a pattern of limited preservation may persist however deep we dig. In some cases, however, it is radically transformed with depth, as it was beneath the Somerset field discussed in the opening chapter, and within the Florida sinkholes that yielded the remarkable pickled brains that figured prominently in an early episode of ancient DNA research.

A striking feature of sites such as these is the permanent water table, the level below which water has never dropped since the archaeological layers were first buried. The whole character of the sediments below this permanent water table is quite distinct from what lies above. Unlike the aerated levels higher up, they have not been mixed or turned by soil animals. There are no earthworms or insects below this permanent water table – indeed, there are none of the soil organisms that need oxygen from the atmosphere. That cuts out the fungi and many bacteria, greatly reducing the pace and extent of decay. Massive fragments, such as pieces of wood, are the most obvious survivors in these pickling sediments, but, on closer inspection, these fragments can be seen to contain seeds, leaves, insects and other invertebrates. A microscope can reveal layer upon layer of intact pollen grains within them. If they were to be fully scrutinized, a single sample of one of these waterlogged sediments could absorb as much of a bio-archaeologist's time and effort as an entire site on which the sediments had been permeated by air.

Even in these submerged layers devoid of air pockets, there are many clues to the organic breakdown and decay that is still going on. One of these clues is the state of the waterlogged wood, and the manner in which it splits and flakes when lifted on to the grassy verge. Another clue is the state of some seeds. Although they were once hard compact objects, they are often found in waterlogged deposits reduced to empty sacs, their outer coating surviving but their innards gone. A third clue is the condition of some of the 'bog-bodies' that have been retrieved from peat deposits in several countries over the centuries. Their skin and their facial features may remain disconcertingly realistic, but many of their internal organs have gone. In most cases, they have lost their brains along with most other internal organs. All these clues point to partial breakdown, some components breaking down, and others not. What is happening is that a select band of anaerobic bacteria (those that can live in the absence of free oxygen) are gradually and incompletely continuing the process of decay. They have a more limited chemical repertoire than their aerobic counterparts, and cannot attack certain types of molecules. It is the tissues that are rich in such resilient molecules that can persist beneath the permanent water table.

The sinkholes at Windover and Little Salt Spring in Florida preserve bodies in much the same way that the Somerset peats preserve the wooden trackways, by excluding most of the oxygen. In fact, there must be something else going on to protect even their brain tissues from decay – the bog bodies of the European peats are not preserved so well. Bill Hauswirth, who began the examination of DNA from the Windover group, suggests that the slight acidity and high mineral content of the water in the sinkhole might contribute to their longevity. The preservation of brain tissues is reflected by preservation of ancient DNA, and the Windover sinkhole has yielded one of North America's large ancient genetic populations. Over 170 bodies have been recovered of which at least 91 retain soft tissue, presumably brain matter, within their skulls.

If we could really dig deep within these waterlogged archaeological deposits, and many of these deposits do go down to a considerable depth, we might expect to encounter a level too inhospitable even for anaerobic bacteria. Such levels would surely provide the greatest wealth of organic residues and molecular survival. Archaeologists have, however, never reached far enough down to find such a level, but they have not been alone in looking. Another group of specialists, the Quaternary scientists, probe even deeper into the Earth's surface to analyse ancient environments and climates. Their deepest probes penetrate the ocean floor.

From the outside, the good ship *JOIDES Resolution* gives little clue to its unusual cargo and crew. That cargo comprises a vast assemblage of pipes, taking up much of the deck. Manoeuvring along the scant space around the stacked pipes is a group of eminent research scientists, whose goal it is to pass those tubes, each one connected to the next, down through the ocean floor. They will go down for hundreds, even thousands of metres, into ocean-bottom sediments that may have taken millions of years to accumulate. The interminable grey sausage they extract and transfer to a plastic tube will be taken to the several floors of laboratories above and below deck. It will provide a wealth of data about our planet's recent history, and the minute fluctuations in its climate while those ocean-bottom sediments were being laid down. If the archaeologists digging on land have failed to reach sediments where biological activity has completely stopped, then surely here within these cores the ocean drillers would find it.

Samples taken from these long grey cores were dispatched to John Parkes's laboratory in Bristol. He found biological activity still present at the ocean bottom, and a few metres into the sediments. This he could show by culturing the samples in sterile conditions, bringing the slow-growing bacteria from this hostile world to life. This microbial activity continued as he sampled further and further down. Astonishingly, there were still viable bacterial moving through their life-cycles at an imperceptible pace, under the immense pressures within the sediments a full kilometre below the ocean floor. If biology was still at work at those remote depths, albeit incredibly slowly, there were certainly no sediments that archaeologists might dig that were sufficiently deep to bring biology to a complete halt. The best we could hope for would always be partial breakdown.

high and dry

Burial does two things. It excludes the larger beasts that have the power to reduce and digest even the densest of bones. Then, further down, with the help of the water table, it cuts off the oxygen supply and restricts the range of decay organisms to particular kinds of specialized bacteria. Oxygen is not the only major supply that can be cut off; many parts of the world are starved of the other major source of life, namely water.

The two most spectacular forms of biological preservation that archaeologists encounter are thus found in extremes of wet and dry.

Arid, water-free sites were to prove a magnet for the molecule hunters, drawing them, for example, to the dry banks of the Nile, where stems of sorghum have been excavated, neatly bound in their sheaths, their seed-heads still intact centuries after burial. The hunters also made their way to the intensely desiccated coast of South America above the Pacific Ocean, where, just under a century ago, the German scholar Max Uhle discovered a series of exceptionally well-preserved human bodies at a beach called Chinchorro. Their internal organs had been carefully removed, the resulting voids stuffed with hair or grass and the incisions stitched up. The stuffed bodies were painted and adorned with wigs. Since Uhle's discovery, several hundred 'Chinchorro mummies' have come to light along that Pacific coast. The intricate mummification practices that preserved them have been traced back as far as 7000 BC, and continued for at least six millennia, providing the first generation of molecular archaeologists with a rich human data-set.

Further inland, conditions are still sufficiently dry to conserve the bodies of animals farmed in the hills. These include the preserved llamas and alpacas to which we shall return in Chapter 6, preserved for over 1,000 years below the house floors of the prehistoric Peruvians who had offered them up in sacrifice. So pristine were the coats of animals buried in this way that the colour and quality of their wool could be studied as if the animal had only just died.

The persistence of such organic remains has not gone on entirely without water, which is never completely absent. All organic materials start by containing at least some water, and even a seed that feels dry may be composed of more than 10 per cent water by weight. Just as there are organisms that can unlock the oxygen from seemingly oxygen-free remains, there are also organisms that can release the water from seemingly dry organic materials. Even in our hottest, driest deserts, decay is not entirely curtailed. Just as John Parkes's deep ocean sediments display one aspect of the resilience of life on earth, another aspect of this resilience can be found within the pans of salt that accumulate on parching desert surfaces.

As those salt deposits form, the highly saline pools from which the water is evaporating do contain life. A bloom of highly salt-tolerant bacteria endows those pools with a reddish hue. They are probably the most salt-tolerant organisms alive, manufacturing their water from long-chain organic molecules. Bill Grant and his colleagues at Leicester set out to look for these specialized organisms beneath the surface, where the salt lakes had turned to solid rock. Deep in the ground, in rock salt laid down 260 million years ago, he found living, reproducing bacteria. They were pursuing their incredibly slow life-cycles imprisoned in tiny vacuoles of hyper-saline solution, in a world as seemingly static and hostile as that encountered by John Parkes a kilometre below the sea-bed.

In comparison with these extremes, the shallow terrestrial sediments into which archaeologists probe will always be relatively hospitable to the organisms of decay. We may excavate objects whose decay has been severely curtailed, but not fully arrested. If this is not much in evidence from a visual inspection of what look like intact objects, it is often betrayed by the state of the molecules within.

burning up and breaking down

Many of the molecules that make up life take the form of a chain. A sequence of smaller molecules, all rather similar to each other, make up the links. The end result is a sizeable molecular filament that can serve a variety of functions. It can be a strengthening device, like proteins such as collagen and keratin, and carbohydrates such as cellulose. It can be one of the cell's molecular 'machine tools', for example the enzyme proteins and RNA. Or it may be like DNA, a filament of coded information, a blueprint from which life's instructions are read. All of these long-chain molecules, or polymers, are vulnerable to the kind of biological breakdown that has been slowed, but not stopped, in the extreme contexts described above. Severely curtailing that biological breakdown is however insufficient in itself to protect the molecules. Even when biological activity is minimal, these long-chain molecules can still end up in a sorry state. This is because biology is not the only thing breaking them down. Chemistry too will do the damage.

The two principal chemical villains are oxidation and hydrolysis. Oxidation eventually leads to a kind of burning up of organic molecules, transforming them into small molecules such as carbon dioxide and water. Where things survive at all, it is because oxidation has been at most a partial process, attacking only the most vulnerable and exposed sections of large biomolecules. Much biomolecular archaeology now involves working with molecules that have been thus disfigured by oxidation. Their analysis entails working back from this molecular debris to the pristine form. The other major chemical villain, hydrolysis, literally means 'breakage by water', and that is precisely what happens to these long-chain molecules. Even very small quantities of water can begin to break the links that make up the chain. In the course of time, the long-chain molecules in dead organic tissue get shorter and shorter. The strengthening proteins lose their strength and elasticity, and the informative DNA loses its code.

What these two major chemical processes have in common is that they are very sensitive to heat. An ancient body exposed in the hot desert sun may superficially look as intact as one buried beneath the glacial ice, but such similarities in appearance are often illusory. The desert body may be brittle and crumbly to the touch, while the ice corpse may retain some of its flexibility, a sign that in the latter the long elastic protein filaments are at least partially intact. Even a few degrees of difference in temperature can make a major impact on the state of biomolecules, because of how temperature affects chemical reactions. Variation in temperature is not one of the factors to which archaeologists paid great heed before biomolecules came on the scene. It is now seen as an important variable affecting preservation. The world's coolest regions, towards the north and south of the globe, and at high altitudes, have a lot to offer biomolecular archaeology.

the plywood principle

Let us look a little closer at the partial nature of biomolecular preservation. Imagine a sheet of plywood. Viewed straight on, it looks like any other sheet of wood – the only pattern being that of its grain. It is only along the edge of the plywood that we can see its layered form, from which its great strength is derived. Now imagine a similar sheet in which all the alternate layers have decayed. Viewed straight on it looks the same. It is only when we flex the sheet and it snaps in two that we discover that its strength has gone.

A number of preserved organic materials are rather like this. Their strength comes from a partnership between two different molecules, minutely intermeshed to confer upon the tissue a strength that neither molecule could achieve alone. Sometimes both those materials will survive the biological and chemical effects of time. More often, one survives rather better than the other.

disappearing bodies

One clear example can be found in the final remains of a human or animal body placed beneath the ground. Skeletons have been central to the search for ancient molecules, partly because they provide access to our own species, and partly because bone is a widespread and familiar archaeological material. It is of one of nature's plywood analogues, deriving its strength from a coupling of distinct molecules, a mineral and a protein. The mineral is the alkaline substance hydroxy-apatite, which is very vulnerable to any acid solution in the immediate vicinity of its burial. The protein is collagen, and this is vulnerable to both chemical and biological attack. A cold, dry, alkaline sediment provides the best medium of preservation, acid sandy soil probably the worst. A burial in the latter may retain nothing except the most resilient part of the skeleton, the teeth, together with a slight stain betraying where the body once rested. Between these two extremes there is a range of possible states. Even within a bone that looks intact from the outside, a host of subtle chemical transformations may have taken place.

These transformations begin as soon as the body dies. The skeleton itself is alkaline, but the flesh around it is mildly acid. This is something with which the living body can cope but, after death, the soft and hard tissues enter into chemical combat. The ancient Egyptians understood this conflict, and one of the first things they would do in preparing a mummy was to take out the large soft organs, conserving them separately.

The combat also has a biological dimension. Soon after death, the body's own recycling processes go into overdrive. Organs such as the liver, kidneys and pancreas have a particularly large number of lysosomes – cellular bodies whose function is to clean up dead tissue that appears within the organ. As these lysosomes begin to break the soft tissue down, the acid juices and enzymes so released eat into the bones, allowing bacteria and fungi to gain a foothold. Once the bone becomes porous in this way, it has lost its first line of defence, a compact exterior, and speedily follows the flesh back into the biosphere.

A seemingly inert mineral object thus retains a considerable vestige of its biological origin. Even the ancient bone's own mineral strength may itself be partly secondary, the result of lime migrating in from the soil, to replace what biology and chemistry have taken away. Other factors may influence its chance of survival: how much flesh was still attached, whether it had been cooked, and how quickly it was buried. The consequent variation in preservation is enormous. Sometimes nothing is left, sometimes the bone survives but is fragile and biscuity, and sometimes it is tough enough to hold. In other cases, however, the processes of preservation can take a body in a quite different direction, when it is more than just the teeth and bones that survives. In places as far apart as the Andes mountains, the lowland peat bogs of north Europe and the permafrost of central Asia, flesh, hair, even internal organs can survive for thousands of years. In each of these cases, some combination of low temperature and the exclusion of water or oxygen has arrested their decay.

The terms 'mummy' and 'mummification' are used for bodies that have been preserved by rapid drying. The words come from the Persian for bitumen, which was sometimes used to preserve dead bodies. What all mummies have in common is that they owe their survival to rapid desiccation. The connective tissues, such as skin, hair, gut and tendon, are the most easily conserved, and they are what many natural mummies are composed of – a surprisingly lifelike hairy hide draped loosely over what survives of the skeleton. The gut is the most durable internal organ, which is why the most frequent internal analysis is of stomach contents and the unfortunate victim's last meal. Other organs, such as muscle, brain and intestine, only survive in the most rapidly desiccated mummies of all.

The very earliest Egyptian mummies were dried by exposure to hot arid winds. By the second millennium BC, mummification had developed into a complex and specialized service catering to the elite. The whole process took around ten weeks, beginning with the removal of internal organs. The emptied corpse was rinsed in wine, then packed with salt. The salt's strong drying effect would eventually produce the stabilized raw material for cosmetic improvement and preservation.

Other societies, as far apart as Australia, China and North and South America, leave similar legacies of desiccated mummies, sometimes through natural drying, at other times with artificial assistance. In the South American Andes, both the natural and the assisted forms of mummification were widespread. The mountain range and its adjacent coastal strip offer a wide spectrum of very arid sites, and a wide range of temperature regimes. Beneath their soils was found a considerable collection of preserved bodies that have formed the basis of a number of molecular analyses. Among them are the South American Chinchorro mummies described earlier. Above them in the Andes, at elevations of 15,000 feet or more, another series of naturally mummified human bodies has been found. Many are around 500 years old and are believed to have been offered as sacrifices according to Inca traditions, to appease the mountain gods after earthquakes, eclipses and droughts. At these high altitudes, preservation is affected not just by dry conditions, but also by low temperatures.

Low-temperature preservation of bodies is found not only at high altitudes, but also at extreme latitudes. Thirty years ago, two brothers stumbled upon this low-temperature effect, while out hunting wild birds in Greenland. They came across the bodies of six women and two children, their flesh and clothing still intact. Subsequent scientific examination revealed the stomach and small and large intestines. The soft tissues were shrivelled but, in one woman at least, the lungs, heart, liver and gall bladder were discernible. They were partially rehydrated and, within the heart, valves, cavities and arteries could be seen. Within the lungs, lenses of soot could be discerned, the consequences of inhaling smoke from the blubber lamp. A date was also established for the bodies. The natural refrigeration of the Greenland environment had conserved these fine details in the flesh of people who had died five centuries earlier.

crumbling wood

Plants display a similar array of modes of survival – from complete disappearance through to uncanny resemblance to the living form, thousands of years after death. While lacking a bony skeleton, many land plants do become woody or 'lignified' in many of their tissues, and it is often this woody tissue that persists, thanks to another 'molecular plywood', one that brings together cellulose and lignin. Cellulose is the substance that gives 'backbone' to big plant and small plant alike. Even a blade of grass would flop without the cellulose shell around each cell. Lignin is the back-up support for really tall plants, and gives them their characteristic woody texture. The combined strength of cellulose and lignin is what allows tree trunks to reach hundreds of metres in height. Rather as with bone, the archaeological wood recovered by archaeologists often has only one member of the molecular partnership in good condition. That is the lignin, which is more resistant than cellulose to biological and chemical attack. In the ancient timber within the Somerset peat, the lignin survives, but the cellulose is gone. Above the permanent water table, soil fungi would also digest the lignin, but they cannot survive without air. Some bacteria can, but are only able to digest the cellulose, which is why it selectively disappears. With only one member of the partnership intact, the strength of the ancient wood has gone. Only the watery matrix is buoying it up. As that water disappears from the ancient timbers that archaeologists bring to the surface, the cells within the wood finally release the strength they have clung on to for millennia under the ground, and begin to collapse. The outer layers of many seeds are composed of a similar cellulose/lignin mix, and these two may be reduced with time to lignin alone, while the starchy interior breaks down even more speedily.

the colour of time

The living world is full of different greens, highlighted with reds, yellows, purples and countless other colours besides, but in the archaeological record all those natural colours merge towards the dark leathery colour of time. In the waterlogged layers of the Somerset Levels, a leaf occasionally retains something of its green hue, birch bark may still be silver, and beetle carapaces may retain their iridescent blue. But like the skin of the bog bodies, the predominant colour is brown. Something similar is encountered in desert sites. The remarkably preserved 4,000-year-old wheat grains from Akhenaten's city, Tel Amarna, may retain most of the physical features of fresh grain, but their colour too has darkened. Fresh wheat grains are light amber in colour. These were like roasted coffee, and other desiccated objects are similarly transformed. In wet and dry deposits alike, organic remains tend gradually to darken.

In the case of food plants, these changes are well understood. The reason that the Egyptian grains look like roasted coffee beans is that a short spell in an oven has a very similar effect on a seed to a very long spell at normal temperatures. In both cases, two of the seed's molecular components, the carbohydrates and proteins, have interacted to produce 'Maillard products', normal and desired products of cooking, that lend aroma, a crust and a dark colour. As well as breaking apart, molecules are in some cases binding together to generate new molecular structures, of which the Maillard products are examples.

Another case of cross-linking is the process we call tanning. When leather is tanned, new bonds are formed between the existing protein chains, giving the whole thing added strength and flexibility. This is achieved by adding tanning agents. Nature also yields its own tanning agents, one of which is released by the mosses that go to make up peat. Many of our bogs are simply an enormous accumulation of old *Sphagnum* moss, piled up upon itself without decay, smothering and pickling other organisms as it develops. One of its breakdown products, sphagnan, is a tanning agent, and its presence causes some of the bog bodies trapped in peat to end up with a fine leather as their skin.

In bogs with a lot of wood rather than moss, a group of compounds called polyphenols can similarly tan organic remains.

Maillard products and tanning processes may help a tissue remain intact, but they are not good news in the search for molecules. All this recombination and cross-linking is a nightmare for the analytical chemist. The hunt for ancient molecules always resembles the proverbial search for needles in a haystack. Science has become remarkably adept at separating chemical needles from chemical hay. They can be persuaded to segregate according to differences in their solubility, their mobility as gases, or their tendency to cling to some other, introduced molecule that acts as a kind of magnet to pick out the needles. These sorting processes have been fashioned into elegant automated procedures that are central to the search. However, the situation is made far more complex when the needles start forming bonds with the hay, at precisely the same time that the internal bonds holding needles and hay intact are themselves weakening.

When the search for molecular information from archaeological sites began, the general assumption was that if archaeological tissue was sufficiently intact that cellular structure could be seen, then intact molecules might be expected. We have learnt to be considerably more circumspect. Organic tissue clearly has a marked ability to retain its visual form, even after the molecules within have greatly changed their form. They may have shortened, oxidized, recombined with other molecules, entered the tissue or left it. In amongst this turmoil, however, highly informative molecules do survive and are the basis of the continuing hunt. Before following the hunt's course, though, we have to address the most basic question of all.

why does any of this survive?

A conflict resides within the picture painted above. It is clear that archaeological deposits, in terms of their molecular dynamics, are nowhere inert. They are far too shallow to elude the profound reach of biological breakdown, in which physical and chemical processes drag large and complex molecules back into nature's melting pot. We have become more, rather than less, aware of the intensity of nature's recycling process. How is it that the ubiquitous forces of degradation and decay fail to recapture these rather shallowly buried remnants of our past? Those remnants are far from consumed. They may be in a poor state, and pose enormous challenges to the molecular archaeologist, but they are there to be recovered. The truth is, we still do not really understand why decay is partial, except that it invariably is. Not that nature is by any means inefficient. Each year, something like 99.9 per cent of all living material is disassembled into small enough molecules to feed a new generation of organisms. That still leaves one part in a thousand that accumulates. In a century, those annual increments have amassed to a tenth of the size of the active biosphere. After a millennium, the accumulation rivals the size of the biosphere itself. On the geological time scale, vast banks of fossil carbon become a major element of the Earth's crust. Many of those accumulating fractions – but by no means all – reside in rather particular deposits, such as those below the permanent water table. Moving back above the waterlogged levels to the active soil below the turf, with its pots, bones, seeds and teeming agents of decay, even here complex molecules survive. They survive within the hearts of the seeds and bones, and in the porous interior surfaces of the fired clay of which the pots are composed. Some would claim that they even survive on the smooth surface of a stone tool. They are there in tiny quantities, and subject to all of the transformational problems outlined in this chapter, but they are there.

At one time, the archaeological record seemed for the most part to be reduced to durable mineral items, nature having long before done its job of recycling the living and organic elements of our world. Subsequently, it seemed there were pockets, culs-de-sac, in which nature had been switched off or excluded, odd places where organic remains could be found in pristine condition. Now it appears that neither of these visions was accurate. Nature and its forces are everywhere, penetrating far further into the Earth's surface than we had imagined. Yet nowhere does Nature complete the job. In all kinds of contexts, some sorts of molecular traces slip out of the grand recycling process and linger on as fading signatures of lost worlds.

5

gaining control

did human evolution stop?

We could be excused for thinking so. Until it comes to the decline of the Neanderthals, archaeology seems to have a lot to say about skulls, brow-ridges, genetics, hominid species' names and the like. After that time, the narrative takes a quite different form, shifting to art and artefacts, burial ritual and monuments, landscapes, social complexity and so on. It seems as if a seismic shift in the human past has left Nature on the far side of a ravine and allowed Culture to take over. An illusion of course, but one with a foundation in a clear pattern in the data. First of all, after the disappearance of our last hominid cousins, it is certainly the case that genetic diversity of the human line was left greatly diminished. Variations in human mitochondrial DNA may be enough to cast light on human migrations, but they are tiny in comparison with what is encountered even among our primate relatives. Two gorillas dwelling in the same small West African woodland could quite easily be separated by a mitochondrial variation beyond that encompassing the entire human race. Within our own species, there is very little genetic variation at all.

Yet while that genetic shrinkage has happened, the cultural diversification among human societies has been phenomenal. There seems to have been an inverse relationship between culture and genes. As the latter narrowed in range, so the former blossomed into an unprecedented variety of forms. The last few thousand years have left behind a greater range of artefacts and artifice than all of the previous 2 million years of *Homo* put together. We are, of course, rather biased in this observation by what survives and what archaeologists have traditionally collected. It would be more accurate to say there has been a blossoming of the more durable things – fired clay, metal, and monuments. For all we know the owners of the first hand-axes were also accomplished wood-carvers, leather-workers and poets, the fruits of whose creativity have failed to survive.

Culture did not really spring from nothing as the first of a series of 'great civilizations' was established, any more than Nature and evolution came to a halt when the global expansion ended and the last fellow hominid disappeared. Nevertheless, at various stages after that time, the material residue of past human societies begins to take on a quite different shape. Large and complex built environments were created within bounded and controlled landscapes. A vast range of raw materials was fashioned into artefacts, sometimes involving the manipulation of fire within carefully constructed kilns.

Alongside all these changes was a shift in the way in which people fed themselves. They changed what they ate and made fundamental changes to how that food was acquired. The way in which many of their food plants and animals grew and reproduced had changed. It was now rigorously controlled within plots in which everything, from the boundaries to the plants within and the soil beneath them, was fixed by human action. For many, the origin of farming lies at the root of all the technical, cultural and social changes that followed. Rather than biological change coming to a halt, the main evolutionary action had passed from our own species to the animals and plants that were maintained within these plots.

Most of the conspicuous civilizations of the last few thousand years have depended upon an increasingly small number of these animals and plants. Prominent among them is a handful of annual plants, whose hard seeds can be softened by grinding or cooking. In the distant past people gathered numerous species of wild fruits, nuts, fish, shellfish, birds and game. In recent millennia their food sources have dwindled to the monotonous expanses of grain crops that still carpet much of the world today. Cultural diversification took place in the context not only of a shrinking genetic range, but also of a dwindling food base. Across the Old World and the New, a series of major civilizations have been built on a narrow range of these cultivated grasses. In east Asia it was rice, further to the west and in Europe it was wheat and barley, in sub-Saharan Africa sorghum and millet, and in America maize. Moving to evolutionary trajectories of the last 10,000 years, our attention has increasingly been drawn from the humans themselves to the plants, in particular to those key seed plants on which many came to depend.

in search of seeds

The rather antiquarian response to the pot of cereal grain recovered from the 1960s excavation described in the opening chapter was quite typical of archaeology at that time. Few archaeologists thirty years ago had a clear idea of what plant tissues survived in the archaeological record, let alone what molecular evidence they might retain. One might think that something the size of an ancient cereal grain would be easily picked out from the excavated soil, but most were missed. Only when they occurred in very dense concentrations were they recovered. The great majority were tossed aside unnoticed. Once archaeologists had embarked upon their concerted effort to find out more about everyday lives and what people were eating, they began to look more closely at the sediments they excavated, passing them through sieves. On dry and sandy sites, it was possible to pass a fair amount of sediment through large screens. This yielded a number of ancient maize cobs from sites in the south-west of the United States and in Central America. However, if the sediments were at all clayey and damp, and if the food plants were small grains rather than large maize cobs, there was very little that sieving could achieve. That all changed, following the simple expedient of mixing the excavated sediments with water.

When mixed with water, what first appears to be an unremarkable sediment, taken from an excavated living floor, hearth or rubbish pit, separates out into two components. Most of the sediment sinks, the mineral part of a soil being over twice as dense as water. A separate fraction rises to the water's surface. It might contain fragments of root, tiny snail shells and the lighter bones of rodents and birds. Most of all it would be composed of dark fragments of plant tissue, blackened pieces of wood, and seeds. The simple process of flotation transformed plant remains from an archaeological oddity to the source of its largest potential data-set. Excavations that were yielding potsherds and bones in their tens of thousands were now yielding ancient plant remains by the million. Here was a form of evidence that could find its way past buildings and potsherds, right through to the prehistoric stomach.

During the 1960s and early 1970s, flotation was carried out in a variety of rather makeshift devices. They acquired names like the 'Siraf machine', the 'Ankara machine' and the 'Cambridge machine', depending on where the device had been put together by a combination of local plumbing and abandoned oil drums, all designed to operate in remote locations with minimal facilities. The aim was to keep pace with the excavation of a site, mixing vast quantities of archaeological dirt with water, in order that the precious 'flot' could be separated, bagged up and labelled, for study back in the lab. I well remember my surprise on seeing under the microscope what had emerged from these dark flots. What to the naked eye were amorphous black pellets, under low magnification revealed structure in astonishing detail. Intricate surface forms were clearly visible, and several separate layers of cellular tissue could be seen through their eroded and broken surfaces. Some oozed a spongy, honeycombed material which provided evidence of the fire that had contributed to their survival, carbonizing and distorting them in the process. At that stage, little attention was paid to their chemical or molecular structure; there was sufficient work to do recording and interpreting the wealth of physical variation in the plant remains. There was enough physical evidence to identify a whole range of ancient food plants among the debris and, by comparison with modern seeds, a wide range of crops was charted through time. In particular the cereals and legumes appeared to survive reasonably well. A category that was even more intriguing than the food plants themselves was the chaff and secondary tissue cleaned from these grain crops and also discarded.

Particularly well preserved were the fragments of seed-head stalk or 'rachis' upon which the ancient cereals had grown. Their form was characteristic to such a degree that they were often easier to identify than the grains themselves. Although these stalk fragments were only a few millimetres long, it was possible to make out where the grains and the surrounding chaff were attached, and how the different sections of the seed-head were linked together. In the most ancient rachis fragments that have been recovered, from wild cereals from sites in Israel, the clean breaks can be seen on each fragment. They are rather like the neat scars left on a branch when leaves fall, a reflection that the seed-head had broken up on ripening, allowing the seed to disperse naturally. On rachis fragments from the last 10,000 years, something quite different is seen. Under the microscope, these natural break points can be seen. The rachis follows a characteristic dog-leg at this point. But the rachis is unbroken here, and instead the untidy breaks are at arbitrary weak points elsewhere. What can be seen in the difference between the two rachis forms is the disablement of a fragmentation pattern fundamental to the natural dispersal of seeds. Looking down the microscope upon flots of small blackened fragments, we can observe the precise genetic modification that transformed wild grasses into cereals and, in doing so, formed the basis for a remarkable range of transformations in human society.

The difference between the brittle rachis that allowed wild grasses to disperse, and the tough rachis that held cereal grains in place and at the mercy of their human predators and guardians, is a fundamental one. The small but visible evolutionary step between the two has intrigued evolutionary scientists ever since Darwin's theory was in print. In the opening chapter of the *Origin of Species*, Darwin concerns himself with the issue of domesticated species, a theme he later developed into a treatise in its own right. Foremost in his thinking was the idea of the 'civilized' breeder who, in contrast with 'savages', could control nature, setting the evolutionary agenda and creating genetic diversity among plants and animals in the human sphere. He had not seen the ancient rachis fragments under the microscope. Neither did he have the archaeological or the genetic understanding we have today. He therefore remained suitably circumspect about how this novel human engagement with evolution had come about, and whether it had happened once, or several times, in one place or in many parts of the world. By the 1920s, the archaeological and genetic pointers that Darwin lacked were available. The question of agricultural origins could be freshly attacked by the Australian archaeologist, Vere Gordon Childe.

Childe was a Marxist as well as an archaeologist, passionately interested in what prehistory could contribute to a grand narrative of the human past, propelled forward in a series of revolutionary episodes. He first developed his notion of a 'Neolithic Revolution' in detail in his classic book, *Man Makes Himself*, published in 1936. In his later volume *What Happened in History*, he described it thus: 'The escape from the impasse of savagery was an economic and scientific revolution that made the participants active partners with nature instead of parasites on nature' (p. 55).

From his extensive knowledge of the archaeological data available at the time, Childe was able to give that general principle some historical substance. He placed particular emphasis on the seeds of two particular grasses, wheat and barley, and on a crescent of land running from the fertile Nile valley in the south, through Israel, south-east Anatolia and Syria, and along the foothills of the Zagros mountains in Iraq towards the west. This Fertile Crescent, bounded by less hospitable environments, was also known to yield early Neolithic sites with ancient potsherds scattered across their surfaces. Although Childe himself was not intent on narrowing his prehistoric revolution down to a single event, he effectively enabled others to do so. He set the stage for them to seek out the very first farms, with the very earliest of these critical grass seeds from which the whole enterprise had sprung.

The spotlight had also been focused upon particular regions of the world by data of a quite different kind. At about the same time as Childe was developing his narrative of prehistoric revolutions, another field was similarly moving forward. The Russian botanist Nicolai Vavilov was hard at work charting the distribution of different varieties of the world's crop plants. He argued that where those varieties were most abundant and diverse, that was where domestication happened. Indeed, those varieties did cluster in regional concentrations, supporting his idea of 'centres of origin'. The Vavilov hypothesis provided a set of world-wide signposts to where archaeologists might find these first farms.

During the Second World War Chicago archaeologist Robert Braidwood started formulating the research strategy that would subsequently lead him to find and excavate pioneer farmsteads at the eastern end of the Fertile Crescent. Up in the Zagros foothills, he found them at such sites as Jarmo. Kathleen Kenyon, from England, would find the ancient grass seeds deep down in her excavations of Jericho, at the western end of the Crescent. Other teams would discover candidates for pioneer agriculture at south-east Anatolian sites such as Çayönü. Before, archaeologists were working with theoretical notions of a great transition. Now, they could examine the actual harvests, the animals and the living floors of the pioneer farmers engaged in the process. Such a proximity with actual lives sharpened the issue of what Childe's 'revolution' meant in human terms.

revolution or evolution?

Childe and many archaeologists after him saw parallels between this prehistoric agricultural revolution and more recent technological revolutions that have similarly had enormous impacts on culture and society. They saw direct parallels between the Neolithic Revolution of prehistory and the Industrial Revolution of more recent times. Just as we can seek out the first steam engine and trace the spread of this radical technological advance, so we could seek out the earliest domestication and trace the ripples of that new idea across the world. This is how Childe's Neolithic Revolution came to be perceived. A challenging natural environment had brought roaming hunter-gatherers together, to where they would make the revolutionary step of controlling plants. This in turn was linked with settling down at a single year-round location where the investment of toil had been made. It also involved some new skills, such as the firing of pots for the cooking and storage of the new foods. In time, the necessary social cohesion was controlled and consolidated in urban sites, linked together in long-distance trade of exotic goods. The settled villages of early farmers had builders, potters and traders who could bring in exotic objects and polished stone. The gatherers and hunters that preceded them inhabited much simpler material worlds. The new artefacts could be plotted across Europe, tracing arrows on the map, starting from the Middle East and eventually reaching to the north and west of Europe.

Childe's revolution was part of a new grand narrative of the human past, a later episode in the sequence that began with the collective improvement of the hominid family, walking upright, expanding the brain and refining the dexterity of the hands. It formed the ecological foundation of the great civilizations that followed. It had a radical origin in space and time, and from that context it spread, replacing the more static and passive societies that remained under nature's control. Much of the archaeological research that first incorporated a search for the food remains was concerned with tracking those origins down, locating the farms where this momentous power over nature was first exercised. By the time this happened, however, a quite different story was being aired.

At the heart of the story related above was the presumed self-evident economic advantage of agriculture over gathering and hunting. Darwin had written with evident disdain about the savages he had met on the *Beagle*'s voyage, and the apparent poverty and brutality of their lives. It was clear to him that the sophisticated breeders of cereals, livestock, pigeons and dahlias in his own society were further up the ladder of life. Childe could still talk of the 'impasse of savagery' a full century after the *Beagle*'s voyage. This Eurocentric view persisted for a considerable time, and it was not until after Childe's death that it was challenged with a more detailed analysis of non-agricultural societies. In 1968, R. B. Lee and I. De Vore's volume *Man the Hunter* presented a picture of affluence beyond the realm of agriculture. The !Kung bushman, for example, achieved an enviable balance between labour and leisure, gathering mongongo nuts that grew wild and forgoing the endless toil of tending plots and controlling the plant's entire life-cycle. The closer scrutiny by anthropologists of surviving hunters and gatherers was drawing attention to their sophisticated and varied interaction with nature, moving them from the status of social fossils to unique societies in their own right. By implication, the first farmers in south-west Asia did not necessarily have an easier time than their ancestors who gathered wild grasses and hunted gazelle. The hard work and narrow diet often affected early farmers' health, and it is far from clear that their yields were any greater than what was available from the wild cereal stands. It is not self-evident that communities on the arrow's path on one of Childe's maps would leap at the prospect of removing entire forests. They may not have been so keen to put aside their seafood, game and fruits of the forest for bread, porridge and the endless backbreaking jobs of digging, planting, tending and weeding.

What is more, as sieving and flotation were generating a rich new data-set of ancient seeds, the new bio-archaeology was not providing an unambiguous picture of early domestication. True, some very early domesticated plants and animals were coming to light along the Fertile Crescent at the root of Childe's arrows, and agriculture in the north and west of Europe was very much later. But here and there early domesticates were being recorded outside this core region, around the shores of the Mediterranean for example. Other regions of the world were revealing their own transitions to agriculture, which, particularly in the case of the New World, had to be independent. Another story might make better sense of all these data. A number of ecologically minded authors favoured another account, one that did not depend upon the notion of a localized technological revolution. Instead, a more widespread and diffuse process of adaptation could account for the data.

dispersed adaptation to a turbulent environment

At the same time as bio-archaeological methods were gaining hold, and the excavations in the Fertile Crescent were proceeding, considerable headway was also being made in understanding how the climate had been changing. This was achieved in large part by examining other kinds of ancient biological remains that reflected climate rather than human activity. What they demonstrated was that the period around the time of early domestication in the Fertile Crescent was one of climatic turbulence. After the glacial maximum, the melting of massive ice sheets around the world was having considerable knock on effects for world-wide temperature, vegetation patterns and animal populations, which shifted and fluctuated rapidly. Humans were not the only species that were faced with adaptation or extinction.

Our recent hominid cousins, the Neanderthals, were, like several largish mammal species, ultimately unable to adapt to the patterns of marked change that accompanied the freezing and thawing of ice on a global scale. Anatomically modern humans were different, their versatility and ability speedily to change their way of living putting them at an advantage in a changing environment. If their own mammalian prey was on the wane, hunters could switch to fishing or gathering – whatever their changed environment offered. Coastal communities across the world were adapting in this way, and coping with change by moving to a broad spectrum of dependable foods. In many parts of the world, such as the wooded margins of the south-west Asian steppes, the gathering of seeds from the herbaceous sward that spread out across the parkland steppe played a major role. Literally hundreds of different species were collected from these prolific natural harvests. In some places, certain grasses were favoured among this mixture.

At a site called Ohalo in Israel, Mordechai Kislev examined a series of blackened plant fragments and found them to be wild barley, 19,000 years old. Along with the seeds were some of the rachis fragments, displaying the clean, natural break that marked them out as wild cereals. He also noticed four rachis fragments lacking that clean break. A further 9,000 years on, at another grass gatherers' site called Netiv Hagdud, over 100 barley rachis fragments out of a few thousand lacked the clean break. In assemblages that are more recent still, all the rachis fragments recovered would lack the clean break. The gatherers of Ohalo and Netiv Hagdud were following an ecological path rather similar to others that had been taken across the world, shifting to dependable, fast turn-around foods in changing environments. The transition from gathering wild grasses to cultivating cereals is not marked by any immediate upheaval. Gatherers and farmers continued to live within walking distance of each other in not dissimilar settlements for 2,000 years or more. The transition was apparently smooth, yet one more adaptation in a species accustomed to constant change.

The pressures of a turbulent global environment and changing vegetation patterns were favouring those communities that introduced some artificial stability into their environments. That was how humans would avoid going the way of other large mammals that failed to adapt to environmental fluctuations. It was ordinary evolution in action, not a historic triumph in the progress of culture over nature but a dispersed and varied set of adaptations, some of which have subsequently led to the world's great crops. It did not have a single point of origin in space or time. It was happening repeatedly and all over the place, and the early generations of environmental stabilizers, if we can call them that, did not necessarily have a better life than their predecessors. It may well have been worse.

Among the proponents of that alternative story was Eric Higgs, a keen advocate of the new bio-archaeology at Cambridge. In the early 1970s he encouraged his team to move their attention from a particular species and a particular time period in the Fertile Crescent. He urged them to look at different places, different times and different species, to find different manifestations of the varied story of human evolutionary adaptation. Indeed, they found there were variants to the story, early cereals and livestock turning up beyond the Fertile Crescent. Other crops were explored by bio-archaeologists in parts of the world remote from this region. Dispersed evolution and multiple origins seemed to offer a better way of accounting for human adaptation than a localized technological revolution. For any of these major resources, we could anticipate multiple routes to agriculture, as different communities experienced similar environmental problems. That was the picture from archaeology. Something quite different was coming from the molecular evidence.

an evolutionary bush

The molecular revolution in biology affected plants as much as it did humans and other animals. Rather similar molecular approaches were taken to rebuilding their family trees or phylogenies. There were some significant differences, some molecules, tissues and sub-cellular structures being found in one but not the other. Nevertheless, plant and animal cells had many similarities in molecular terms. In each case, their tissues were largely made up of proteins, carbohydrates and lipids. The genetic blueprints of plants and animals alike were stored in the form of DNA, both in the cell's nucleus and within some of the other structures inside the cell. Just as with human genetics, the molecular revolution in crop plants first gained momentum with proteins rather than with DNA. Just as the blood proteins gave an early molecular insight into human genetics and evolution, so seed storage proteins and a group of enzymes called isozymes did the same for crop plant genetics. Minor variations between the different proteins could be used to build a phylogenetic tree for different varieties of any particular crop species. The shape and pattern within those trees could reveal to us aspects of how the past had unfolded, with a direct bearing on the contrasting models put forward by Childe and Higgs. However, a closer look at these phylogenetic trees calls into question whether 'tree' is the right way to describe them.

Returning for a moment to human evolution and the Neanderthal story, we were brought into contact with Mitochondrial Eve and with the family tree that springs from her. That phylogeny was more of a bush than a tree. There was no main trunk, no lofty canopy. The whole thing was a busy cluster of branches, spreading laterally rather than reaching vertically. The shape of the bush was traced by working back from the DNA sequences of living women, without prioritizing or ranking one above the other, but simply trying to make sense of the similarities and differences between them. Its form was traced by projecting backwards in time, progressively joining the shoots occupied by modern samples into minor then major branches and finally to a common stem. We need further to imagine that the bush was linked to a series of neighbouring bushes by underground stems. Those underground stems may be followed to such neighbours, allowing us to view the bush in the context of a much larger picture. In the case of the 'human bush' stemming from Mitochondrial Eve, some related primate such as the chimpanzee or orang-utan provides a suitable neighbour through which it may be 'rooted'. In the cases of the plant species involved in agriculture, they too have their own evolutionary bushes, rooted in relation to their own set of relatives.

The great thing about the complex pattern of prolific branching is its wealth of informative content. If we look at the form of a real bush, we learn a great deal about the history of its development. It may have grown up in a fertile environment, every branch radiating out to yield countless shoots. Or it may have been caught without water or in the shade at some point, and bear the scars of withered and dead branches to prove it. From its shady position, an odd cluster may have reached through a gap to find a shaft of light, and yielded a fresh branching cluster of young growth. All this is in the bush's past, and yet is preserved in the patterns of its final growth form.

Much the same is true of the evolutionary bushes that grow in the virtual environment of an evolutionary biologist's computer. The pattern of their branches mirrors what happened in a species' distant past. Nodes of prolific branching and radiation recall times when selective pressures were eased, just as a narrowing of the branching pattern can reveal the reverse, when only a few lines passed through a genetic bottleneck. The term 'bottleneck' refers to some geographic or demographic constriction that filtered out many of the branches, just as the heavy shade around a single shaft of light accommodated only a few branches of our living bush. These branching points and bottlenecks that we only discover through analysing similarities and building evolutionary bushes serve as virtual fossils of ancient paths in evolutionary history.

Such information is gathered by reference to characteristic lineages tracing a sequence of haplotypes, the genetic 'surnames' introduced in Chapter 3, in the context of Mitochondrial Eve. All species generate a range of haplotypes for study. In the case of domesticated animals, those haplotypes may be extremely similar to the haplotypes studied within humans. In more distantly related species such as plants, they will be distinct, but the phylogenetic analysis of the form and structure of their evolutionary bush will remain similar.

branching patterns and agricultural origins

Let us consider the two extreme versions of how agriculture might have come about, and how they will affect the form of evolutionary bush generated from the crop species involved. If we start by building that bush from all the wild relatives of that crop living today, then those relatives will form the many tips of the bush's branches. Beneath those tips, the bush's structure and root will be assembled by computer analysis. When it comes to adding the domesticated forms, if we take Childe's model to its extreme, then all the domesticates will spring from a tiny portion of the evolutionary bush's crown. This corresponds to the wild forms growing in the specific location of agriculture's origin. If we take Higgs's model to the extreme, the bush will look quite different. Domesticated forms will be peppered across the crown, springing from several branches and reflecting the many locations of the transition.

These bushes could also be constructed from the protein data that became available before the DNA evidence was sufficiently well studied. As that protein data accrued for different crops, so the pattern they produced repeatedly fell closer to Childe than to Higgs. The protein diversity among the wild plants was great, giving the phylogenetic bush a broad crown. When domesticates were added to the bush, their own diversity was seen to be much narrower. They sprang from a small part of that crown. Those patterns on the constructed bush could be transferred to a map, according to where the wild varieties grew. In the case of maize, for example, wild maize plants are found in various parts of Central America. The protein patterns in domesticated forms, however, relate them all to a single line, which grows today in the Jalisco Valley, not far from Mexico City.

As DNA methods came on stream, something similar could be done with the molecules at the heart of life. A recent example is the work of Manfredd Heun at the Agricultural University of Norway. He looked at the DNA of one of the most ancient wheat species around. It is a slender plant called einkorn wheat, which is nowadays rare, but was an important crop in early prehistory. From bio-archaeological evidence, it appears to be one of the first, perhaps the very first, plant species displaying physical evidence of domestication. Heun compared a range of modern wild and domesticated einkorn wheat samples using a method not dissimilar from the DNA fingerprinting employed by the police to catch criminals. This method is different from the DNA analyses featured in other chapters, that target and work with a single stretch of DNA sequence, bounded by carefully designed PCR primers. In Heun's method, as in DNA fingerprinting, restriction enzymes are used to fragment the total body of DNA into many strands of various lengths. Any pattern within this constellation of lengths is a direct reflection of where along the DNA strands the chosen enzymes have been able to bind to and break them. A convenient way to reveal this pattern is to allow the lengths to migrate and separate along a suitable gel which, when stained, produces the characteristic 'fingerprint' by which one wheat variety, or indeed one criminal, is identified from many others.

The technical term for the patterns studied by Heun is 'amplified fragment length polymorphisms', normally abbreviated to AFLP. Heun established the AFLP patterns for a range of stands of wild einkorn from Turkey, Syria and neighbouring countries. He then added selected populations of domesticated einkorn to the analysis. When the different patterns were grouped according to the respective similarities, whatever statistical method was used brought the domesticated einkorns very close to one particular population of wild einkorns.

This wild einkorn population was from the slopes of the Karagadag mountains in south-east Turkey. No more than 100 km from these hills, a Dutch archaeological team had been digging a very ancient settlement at a place called Çayönü, an early fixed settlement that had yielded some of the earliest carbonized einkorn grains. So here, within walking distance, was a contender for the earliest domesticated cereal deposit in the world, and a stand of wild cereals that could be the ancestor of all domesticated einkorn. Heun's final words in a paper in *Science* (1997, 278: 1314) were suitably circumspect:

> Localization of the precise domestication site of one primary crop does not imply that the human group living there at the end of the Palaeolithic played a role in establishing agriculture in the Near East. Nevertheless, it has been hypothesized that one single human group may have domesticated all primary crops of the region.

In the same issue of the journal, Jared Diamond was keen to push the boat out rather further:

> [A] long straight line runs through world history, from those first domesticates at the Karagadag mountains and elsewhere in the Fertile Crescent, to the 'guns, germs and steel' by which European colonists in modern times destroyed so many native societies of other continents. (1997, 278: 1244)

Even *The Economist* (15 November 1997: 127) was moved by that ancient business opportunity emerging from the ruins of Çayönü and its neighbours:

[T]hese sites really are the earliest evidence of agriculture, rather than merely the earliest that archaeologists have yet discovered. It also suggests that the people who built them were the most important inventors in history. The West certainly owes its existence to them. So in all probability do the civilizations of India. Even China is suspected by some to have drawn its inspiration from what they started in the Middle East. And we do not even know their names.

Not just in the press, but also in the research community, the molecular evidence stood in contrast to Higgs's idea of a widely dispersed adaptation to a radical event at a particular location, or at least a very small number of such locations across the world. Although dispersed evolution was more in keeping with modern ecological thinking, the idea of a great step forward in the human story was making a comeback. Furthermore, the various early European dates outside the Fertile Crescent had fallen by the wayside, victims either of flawed dating or of flawed identification. The acceptable carbon dates seemed instead to radiate out from the Crescent.

It seemed clear where the molecular evidence was going, but not as clear how much the archaeological evidence was in support. In general terms, the archaeological chronology worked. It had in any case been used in the genetic argument. However, the fine detail of the archaeological record revealed a picture that was more complex than the simple model of radical origin and dramatic spread. First of all, it proceeded at an imperceptible pace. A thousand years after the first domesticated crops appeared within the first generation of farming settlements, the Near Eastern landscape was far from transformed, and hunter-gatherer communities were still widespread. The new approach to food had not yet begun to spread across Europe. When it did, its spread was surprisingly discreet, with negligible impact upon the woodland covering much of Europe for several thousand years. Wild resources continued to be hunted and gathered even by those tending cereal plots. The whole process seemed slow, bringing to mind gradual adaptation rather than revolutionary change.

To bring the explanations coming from archaeology and genetics closer together, various researchers have attempted to look at the genetics of archaeological specimens, by amplifying ancient DNA from the battered plant fragments that were increasingly being recovered from archaeological sites.

the ancient molecules

It still surprises me how uninquisitive we can be about aspects of what we observe that lie beyond the realm of our immediate questions. By the time Terry Brown and I had been drawn together by our common curiosity about the survival of DNA in ancient plants, I had been studying blackened fragments under the microscope for almost two decades, drawing them, measuring and counting them and recording cells, hairs and other features. When Terry posed an innocent and perfectly reasonable question for a molecular biologist – what are they made of? – I surprised myself by how speedily I exhausted my rather flimsy knowledge. I suggested to him that they were largely composed of carbon, that they had been exposed to a reducing fire, and that the steam released had generated the bubbly, honeycomb structure and the black residue. However, I knew full well that so-called 'carbonized seeds' come in various shades of dark brown and black, and by no means all display the honeycomb structure. We archaeobotanists had shown little interest in even the most basic of chemical assays to check this out, and remained happy to continue observing, drawing, counting and so on. Much the same was true of other forms of seed preservation. Seeds from both waterlogged and desiccated deposits survived by virtue of the cessation of biological activity in the absence of air or water – another over-simplification. Mineralized seeds were a kind of fast fossil, in which a calcium salt in the surrounding matrix had replaced much of the decaying plant tissue.

It is now much clearer how we have over-simplified each of these categories. Like other bio-archaeological materials, the majority of ancient plant remains persist after the routes to decay and breakdown have been partially, rather than fully, arrested. It is not easy to completely reduce a plant fragment to carbon. Charcoal burners build very specific fires to achieve this. Nor is it easy fully to exclude either oxygen or water. The oxygen in a waterlogged deposit may be extremely low, but such deposits rarely offer a complete barrier to diffusion. Similarly, a 'desiccated' seed may retain a surprisingly large quantity of water in its make-up. Mineralization is by its very nature a partial process. So we have many plant remains in which molecular breakdown has been considerably reduced rather than stopped, and sufficiently so for their visual form to be studied in detail. Their molecular preservation was likely to be varied, and their DNA content was difficult to predict.

We assembled ancient cereal grains in a variety of different preservation contexts. Desiccated wheat grains excavated from the Jordanian desert seemed the best bet, and some of these were also partly mineralized. We were assembling these soon after the publication of Miocene *Magnolia* and the problems of water were not widely appreciated. So the waterlogged wheat grains we recovered from a mediaeval well also seemed a likely source. Finally, the most ubiquitous but least promising category, carbonized material, was an unlikely source. Even if they were only partially exposed to fire, the heat would surely destroy the DNA. However, these blackened seeds had emerged as the core of the fast-growing archaeobotanical database around the world. Brown built a probe that was specific to wheat DNA, and which carried a radioactive marker. If any of the sample retained wheat DNA, the marker would bind and emit sufficient radiation to mark a photographic film with a spot. Both the desiccated and waterlogged wheats came up positive; some of their DNA had survived. We were taken by surprise by the carbonized seeds, which also came up positive.

For our sample of carbonized seeds, we had chosen a rich cache of cereals from an underground pit at the prehistoric hillfort of Danebury in southern England. By this time, we had many specimens from which to choose. It was a chance selection of numerous assemblages that was taking up rows and rows of storage in our Cambridge archive. With hindsight, we can estimate that at least 80 per cent of the assemblages in that archive would have completely lost their DNA. Purely by chance, we chose one that had not. It would have been perfectly in order to discard carbonized seeds after that first negative in the pilot study, but it just so happened we had chosen the one assemblage in six that encouraged us to continue. It was now possible to go to the heart of the domestication issue in search of DNA.

a suitable blueprint

The ancient DNA research on humans had made much use of Anderson's sequencing of the human mitochondrial genome. Food plants also have mitochondrial genomes, but the same intense effort that has been applied to sequencing our own DNA has not been lavished on the very many plant species we use for food. There are now some detailed mitochondrial maps for plants, which may figure more prominently in future studies. In the meantime, there is another structure within the plant cell that can be explored in a similar way. While mitochondria are the cell's powerhouses, burning up energy to fuel the cell's activities, the task of the chloroplasts is to capture that energy from the sun. Just as the cell has several mitochondria, it also has several of these bright green chloroplasts. Just like the mitochondria, they also carry a circular DNA molecule, and are normally inherited through the maternal line.

The chloroplast genome in plants is considerably larger than that of the mammalian mitochondrion, and is highly conserved across the plant kingdom. While it has been used to explore the ancient evolutionary origin of the plant kingdom as a whole, it has been less widely used in the kind of micro-evolutionary story of interest in the human past. For that, a greater focus has been placed upon the chromosomes within the cell's nucleus. In most plant species, the nuclear sequence is only patchily known, but certain genetic regions are very well charted. These tend to be around genes with some economic significance, such as those affecting yield, disease resistance or cooking qualities. At either end of the genes are the long non-coding regions that occupy so much of the nuclear genome, the so-called spacer regions. The punctuation marks within the DNA sequence that signal the start and finish of the reading of an individual gene are referred to as the promoter and terminator sequences. The non-coding DNA is not confined to the stretches that lie outside the genes themselves. At various points on a gene, lengths of DNA sequence are apparently skipped over or edited out when it comes to actually assembling the sequence of amino acids within a protein. They are called 'introns'.

Hence we have a number of different kinds of sequence within and around the gene itself – spacer region, promoter and terminator sequence, intron – that are separated from the eventual design of proteins. We assume therefore that they are also separate from the direct impact of the forces of external, whole-organism selection. Instead, they serve as valuable accumulators of neutral evolutionary mutation, recorders of evolutionary history. Various of these have been brought together to probe into the ancestry of nuclear genes.

agriculture in the new world

It was back in Allan Wilson's Berkeley lab, the home of so many innovations in this field, that the first attempt to amplify ancient DNA from a crop plant was under way in the early 1990s. The crop in question was the prime example of the transformations that prehistoric breeders could effect. The bulky maize cobs that end up on our table bear little obvious resemblance to the slender-headed wild ancestor from which they were bred. In genetic terms, however, domesticated maize is so close to its wild ancestor, 'teosinte', that many experts regard them as members of a single species. Yet in a few thousand years, domestication has turned a modest grass head into the high-yielding monster we feed on today.

Maize is a plant of enormous diversity. The cobs and kernels each vary considerably in size, shape, colour and texture. In the modern period, North American maize has been modernized and streamlined into a select suite of successful varieties, but it is still possible in the rural parts of South America to find an extensive and colourful diversity of forms. This variability has led some to query the notion that in the New World, like the Old, crop domestication had a point of origin in space and time. Perhaps a scattered origin would better account for the regional diversity, especially in the south. However, a mounting body of genetic and archaeological evidence has been assembled to support a single point of origin. The genetic evidence related in the first instance to isozymes. The protein variation among domesticated maize was narrower than among wild teosinte, and tallied with particular populations of teosinte growing today in the Jalisco Valley on the west coast of central Mexico.

At roughly the same latitude on the east coast of Mexico, Richard MacNeish mounted one of the pioneering archaeological explorations of early agriculture in central Mexico. His team worked through the deep deposits of the rock shelters visited in prehistory along the Tehuacan valley. Within these were some of the earliest levels containing maize cobs. A whole series of cobs in different shapes and sizes was recovered, ranging from diminutive forms much closer in scale to that of the wild teosinte, through to the substantial form of cob more familiar today. At first, the time scale of agricultural beginnings in the New World appeared to have an antiquity similar to the Old, but refined carbon dating has brought the dates forward by several thousand years. Nevertheless, a time sequence could still be constructed beginning in central Mexico, close to the most similar populations of wild teosinte today. Here domestication began and successive generations of cob swelled to their familiar enlarged size. The carbon dates then trace the spread of maize farming north and south, where, alongside beans and squash, it went on to fuel some of the best-known urban civilizations of pre-contact America. The single-origin model seemed to fit, but not completely. A few genetic points needed tidying up.

The diversity of maize was not just related to physical appearance. If it were, it would amount to a weak source of evidence, as so little is known about the relationship between physical and genetic diversity. It also related to the DNA sequence. Some sequences that had been studied argued against a bottleneck in the last few thousand years. One such sequence was the promoter region of gene associated with the metabolism of alcohol within the plant, a gene labelled *adh2*. Enough is known about these genes across the grass family to make an estimate of mutation rates within, in other words to calibrate its molecular clock. For this sequence, the average divergence rate of two individual plants was calculated at 1.6 per cent per million years. To put this another way, imagine gathering together all the modern progeny from the very first kernel planted by the very first American farmer. Among the vast assemblage of modern kernels so gathered, the sequence of any selected pair of kernels would deviate from one another along this sequence by an average of 0.008 per cent. The deviation observed by the Berkeley team was greater, by several orders of magnitude. They measured it at 2.2 per cent. Even if we allow a few more farmers to scatter several kernels, the breadth of genetic variation seemed large for a single event.

As the genetic consensus was for a single domestication event, some geneticists had suggested this might have to do with a molecular clock that accelerated after domestication. The estimate of a convergence well over a million years back needed to be reduced accordingly. Even if this were far-fetched, we have already seen how the precision of the clock is being sorely pushed at this level of resolution, and is in need of corroboration by the far more secure dates from archaeological specimens.

Pierre Goloubinoff, a graduate student of Wilson and Pääbo, set about assembling some suitable archaeological specimens of early maize, with a particular emphasis on South America, where such striking diversity is seen today. One group of kernels was from the north coast of Chile. They had been added to a human burial 1,500 years ago, and survived in a state of desiccation. Another group came from Christine Hastorf's excavations of a Wanka settlement in the mountains of Peru. They were from around the time of European contact, and were carbonized rather than desiccated. A much older cob, whose mode of preservation was not specified, came from the north coast of Peru. At 5,000 years old, it was significantly closer to the earliest maize cultivation than to the modern epoch.

Goloubinoff amplified a 315-base-pair sequence from the target sequence in a sample from each of these ancient specimens, together with modern specimens of both domesticated maize and its wild ancestor, teosinte. All three ancient specimens yielded a DNA product, including the carbonized sample, adding support to Terry Brown's findings. Two results were of particular interest. Judging from these results, the full range of domesticates did not narrow down to the particular populations of wild teosinte inferred from the isozyme evidence. Second, the 'speedy clock' argument did not work. If the molecular clock had speeded up as some had suggested, then the variation among ancient specimens would be considerably less than among modern samples. What Goloubinoff found was the reverse – it was almost 30 per cent greater. The molecular clock was not running fast, and the youngest common ancestor for the ancient and modern maize cobs was placed millions rather than thousands of years back in time.

The results do not actually dispose of the single-origin model. They simply demonstrate that the domestication of maize is not marked by a severe genetic bottleneck. Multiple events of some kind have retained many lines of wild teosinte within the gene pool of domesticated maize. There is a variety of scenarios to explore. The first American farmers could have started with varieties of wild teosinte significantly more diverse than those that survive today. Alternatively, they could have picked up the diversity from wild pollen as they journeyed south, although they would have soon got past the distributions of wild teosinte. The third possibility was that maize was brought into domestication, not once, but many times, and perhaps in more than one place.

This work was done in the early days of ancient DNA science, and a number of refinements could now be added to narrow down the possibilities. These days we might tighten up the research design. A further look at geographical variations would help. The wild pollen was a problem, especially as maize and teosinte are cross-pollinating plants that freely release their pollen to the air. It might be interesting to examine the immobile female line through the chloroplast DNA. This has been done on modern maize varieties, and has tended to confirm that the maternal lines in domesticated maize incorporate a wide range of wild teosinte lines. Something we cannot do with maize, but can do with several of the other grass species that have become major cereal resources, is to focus upon self-pollinating plants that release little pollen to the wind. These various refinements have been brought together on another ancient DNA project on the far side of the globe from Mexico.

another fertile crescent in the east

Richard MacNeish was guided in his search for early maize cobs by Nicolai Vavilov's approach, using the distribution of living relatives of the modern crop. This took him to Mexico to search for early sites within the centre of diversity for maize and beans. In the Old World, the equivalent centre of diversity for wheat, barley and a range of other useful plants and animals took Robert Braidwood to the Zagros mountains in the Fertile Crescent of south-west Asia. Further to the east lay another fertile crescent – another centre of crop plant diversity. This 'Eastern Fertile Crescent' took in a series of high plateaux at the south-eastern end of the great Himalayan range. Starting in the west around Assam, it stretched through Myanmar (Burma), northern Thailand and Laos, and into the southern provinces of China, Yunnan and Guizhou. This is the realm of that other major world crop, rice. It is here that the wild relatives of rice flourish in their greatest diversity, sometimes in the high plateau marshlands reminiscent of the paddy fields that imitate them.

The Eastern Fertile Crescent became the target of a number of explorations of the eastern beginnings of agriculture, particularly in India and Thailand, the areas most accessible to archaeologists from the west. Early dates were published, stretching back as far as the seventh millennium BC. Somewhere here, perhaps, the original rice farmers could be tracked down. Rather as with the early American farming sites, the early dates failed to stand up to close scrutiny. In time, they were adjusted or discarded, and the antiquity of rice farming within the Eastern Fertile Crescent has now shrunk to less than 5,000 years. At the same time as these dates were being put in order, news was arriving from China that changed the whole story.

Much earlier sites, rich in carbonized rice grains, chaff and straw, were coming to light well outside the Eastern Fertile Crescent. In the lower reaches of the Yangtze River, sites such as Luojiajiao and Hemudu were yielding rice that was 7,000 years old. By the late 1980s, even earlier sites had been discovered, with vast quantities of rice preserved in a waterlogged state. Settlements at Pengtoushan and Jaihu were harvesting rice 8,000–9,000 years ago.

Even further outside the Eastern Fertile Crescent, a rice geneticist working in the city of Shizuoka in Japan had been taking an interest in these early sites from the Yangtze River. Before constraints on movement around China relaxed, Yo-Icho Sato's main contact with these new discoveries was by word of mouth from Chinese graduate students who had travelled to work with him in Japan. It was still not politically easy for Sato to go to the sites to sample for himself, but his students were able to bring specimens of the ancient rice back to his lab in Shizuoka. After a while he had assembled enough ancient samples to explore their genetic relationship with the better documented material from further south. By the early 1990s, he had read the reports of ancient DNA findings published by Russell Higuchi and Ed Golenberg, and was keen to try them out.

The first thing he noticed, simply from microscopic examination, was that some of the ancient rice from Hemudu was wild. Vavilov had been building his argument before it was generally realized how much climate, and the vegetation patterns dependent on it, had changed over the millennia. These Hemudu rice grains provided direct evidence of that fact. Wild rices had clearly changed their distribution over the last few thousand years. Careful survey of the modern flora has now yielded wild rice, found growing some way outside the Crescent, though we still have to factor in climate change to push the distribution far enough to the north to encompass the earliest sites.

Sato turned to the chloroplast genome, which in rice could be split into two groups, according to a stretch of DNA that was sometimes missing. Sixty-nine base-pairs found in the chloroplast sequence in some types of rice were completely absent in others. This mutational accident, known as a 'deletion', happened far back in the ancestry of the rice. This is clear from the fact that many wild rices also carry this deletion. It is faithfully reproduced from generation to generation, and so now serves as a lineage marker.

Sato's trial run on ancient rice DNA was carried out on a 1,200-year-old sample from Japan. The DNA was remarkably well preserved, with strands of over 1,000 base-pairs amplified, and he was encouraged to move on to his precious collection of very ancient Chinese rice grains. The results he gained from these displayed a clear contrast between the middle Yangtze valley to the north, and the Eastern Fertile Crescent to the south. In the wild and domestic rice alike, the 69-base-pair deletion proved to be absent from the Yangtze specimens. When these sites were occupied, the climate had been a few degrees warmer, enough for perennial rice varieties to spread further north than the area of their current distribution. Perhaps it was on the edge of that distribution that significant changes took place. Here, where the viability of wild rice lessened, so it was brought into domestication, switching from perennial to annual habit. There is no sign of a genetic path leading to the deletion-carrying lineages, which Sato argues derive from the annual wild rices to the south. Here was a well-argued case for multiple domestication of a species, in locations separated by 1,000 kilometres or more.

a return to the cradle

Maize and rice are crops of global importance and each has a long prehistory. However, their past in each case remains significantly shorter than in the region to which Childe drew attention, and which has been thought of as the cradle of civilization for many generations. This is the Near East, in the region described as the Fertile Crescent. It may be that future discoveries may take rice agriculture that little bit further back in time to make up the difference, but they have not yet done so. For many, the Fertile Crescent 10,000 years ago remains where it all started, and at some point within that region are those original pioneer settlements, whose occupants changed the course of history forever.

Many excavations have now been conducted along the Fertile Crescent, especially along its western flank. The changing political fortunes of Iraq and Iran have made it difficult to follow up the work of Braidwood's team along the eastern flank, but early sites are now known from Syria, Turkey, Lebanon, Israel and Jordan. There have recently been two principal contenders for the primary place in the transition to agriculture. Some researchers have focused their attention upon a cluster of four sites in the southern part of the Jordan valley, of which the best known is Jericho. The four sites are very close together, and may possibly have domesticated wheat and barley more than 10,000 years ago. Others have argued that the cereals from that date were wild, and that a group of sites 700 kilometres to the north in south-eastern Turkey have more securely identified early domesticates. What is not in question is a key series of founder crops that were domesticated in that very first episode of agriculture: two species of wheat – einkorn wheat and emmer wheat – one of barley, and a small group of legumes. The key to really getting to grips with the single versus multiple origins debate was how to understand these founder crops in the very first stages of domestication.

Terry Brown's group had been working with wheat from the outset of the ancient DNA project. Since their amplification of DNA from the carbonized wheats from Danebury, they had moved on to explore other carbonized deposits of even older dates and from elsewhere in Europe. One of these was a packed grain store in Greece that had burnt down 3,300 years ago, leaving great caches of carbonized grain intact within their storage bins.

One of their target sequences was a gene group responsible for a series of proteins called the high molecular weight (HMW) glutenins. These are proteins formed in the shape of a flexible spring. Our modern 'bouncy' loaves owe their texture to this springy protein. Because of their importance in baking, they had been well studied by modern dieticians, and were known to display the kind of variation that might provide a clue as to their recent evolutionary history. Each species of wheat has a certain number of glutenin gene loci, each occupied by a certain version or 'allele' of the gene. Some of these alleles no longer do any useful work. Mutations have disabled them, turning them into 'pseudogenes' carried like dead weight from generation to generation. Others have acquired mutations that slightly vary the shape or length of the molecular 'spring' they are building, but not in a way that damages the plants. Among these are alleles that make particularly well-formed springs, not in terms of the health of the plant, but of the interests of the baker in making light and springy bread. All this variety is the raw material from which evolutionary stories may be assembled and narrated.

One interesting finding was of the bread-baking allele. This was the one that optimized the dough's ability to retain large bubbles of carbon dioxide and produce a light elastic loaf. Modern bakers have certainly selected for this allele, though for how long is uncertain. We are not sure how the artificial selection for any such character was effected prior to the nineteenth century. Carbonized loaves such as those from Roman Pompeii show that leavened bread was made, though not necessarily with specially selected flour. After examining ancient DNA, we can now say with confidence that Greek bakers had already selected for the very same genes that modern bakers favour to produce a light and springy loaf. They had done so well over 1,000 years before Pompeii was buried. Not only were prehistoric communities selecting for bigger and easier harvests, but they were also systematically targeting attributes invisible in the field that required a clear understanding of the notion of a genetic line.

The glutenin genes had yet more things to tell about the past. One particular aspect of the gene caught the attention of Robin Allaby, one of Brown's team at Manchester. At one end of the gene, the promoter region, the sequence varied quite significantly. Using the molecular clock estimation, he reasoned that the different lines had diverged from one another 1–2 million years ago, in other words, a very long time before agriculture had begun. This in itself is not surprising – the genome of any taxon preserves quite a number of ancient legacies such as this. Allaby was interested in treating the different phylogenetic branches from the split as markers, to try to make out patterns of evolution and dispersal in the crop. A starting point was to look at a little known wheat, which was once at the heart of Old World agriculture – a species called emmer wheat.

Although so important to prehistoric farmers, this particular species of wheat is now a rare crop. The last 1,500 years have seen its progressive replacement by modern bread wheat. Twentieth-century agricultural improvement could well have wiped it out, were it not for the recent move to conserve genetic diversity. The scattered emmer wheat populations that survived have been sampled and stored with two major cereal archives at Gatersleben in Germany and Aleppo in Syria. Allaby worked his way through these archives to chart the patterns within the promoter regions of their glutenin genes. A very clear pattern emerged.

His population of wheats clearly subdivided into two distinct branches, following an evolutionary split some 1–2 million years ago. He labelled the two branches from this split the alpha and beta clades. The alpha clade was the more common and the more widespread, found in emmers throughout the world. By comparison, the beta clade was far more localized. It was encountered in sites in the eastern Mediterranean running into central Europe. When he turned his attention to the wild precursors to domesticated emmer wheat, found today in various parts of the Fertile Crescent, another pattern emerged. The beta clade was quite common towards the Anatolian end of the Crescent, where those early farming sites that attracted Heun's attention are to be found. However, the alpha clade peaks further to the south, in Israel. This is where a second series of early farming settlements has been found, also with some very early radiocarbon dates for crop domestication. Taking the archaeological and genetic evidence together, it seems that a very similar sequence of events happened at least twice, the same key transition in human ecology, and propelling two very similar spreads of the new crop. In the longer term, one came to predominate across the world, but even so, a substantial genetic trace of the other has survived. There could quite conceivably have been several more such trajectories, one overlaid upon another, of which these two happen to show up on this particular stretch of the DNA sequence.

an emerging picture

There have been two ways of seeing and interpreting this major turning point in the human past. The first of these has emphasized isolated revolutionary events, a radical change of direction, rather like a major scientific invention of more recent times. The second has instead emphasized dispersed gradual processes in response to similar global pressures, the convergent evolution of communities steered down the same path of adaptation to variations in nature. Underlying these differences was a wide spectrum of views about the nature of human history and progress. At one end of the spectrum was the view that humans were on a path of improvement by historical action and unique invention. At the other end was the view that we, as much as other species in nature, were subject to the rough and tumble of global fluctuation and natural selection. In that view, the transition to agriculture was just one facet of our adaptation to that process.

The first interpretation has an affinity with a single isolated origin and an outward spread from there, from time to time triggering secondary domestications. By contrast, the second interpretation would have an affinity with widespread multiple domestication. Such genetic, molecular and archaeological evidence as we now have fits comfortably with neither extreme. The most parsimonious way of accounting for it is as a transition that is rather patchy and bunched, both in space and in time. Some species enter domestication through the kind of tight genetic bottleneck that would suggest a very restricted event, but others do not. In the Neanderthal debate of Chapter 3, the ancient DNA was considered in relation to contrasting hypotheses that were also rooted in deep-seated notions about human history and progress. In that instance, the DNA came down on one side of the argument. With the transition to agriculture, a more diverse picture comes into view. The fine detail of the living past, rather than supporting one or other unequivocal pattern, is pointing up the diversity and contingency of past life. When we project back from the diversity of the present to find its roots in the distant past, the simple explanations are the favoured ones. When, however, we inject a rich and detailed body of evidence from the past itself, perhaps it is not surprising that those simple explanations become less persuasive, and the past becomes as diverse and intricate as the present.

6

ending the chase

the roots of captivity

The stretch of land that came to be known as the Fertile Crescent had a double attraction for those in search of agricultural origins. The arc of land that reached from Israel in the west, through south-east Turkey and Syria to the north, and on to the slopes of the Zagros mountains in Iran to the east, was not just the home of the wild ancestors of so many major cereals and legumes. Roaming through the woodlands in the north and west of the crescent were the wild boar and the massive aurochsen that gave rise to domestic pigs and cattle. Higher in the foothills and mountains to the east, two agile grazers, the mouflon and bezoar, dodged their predators. These were the wild ancestors of our modern sheep and goats. Domestication was not just about plants, but about animals too. When, 10,000 years ago, hunter-gatherers made their way across the Fertile Crescent, they came across both the plants and the animals, living in the wild, that would subsequently feed the greater part of the world's population.

It was the animals in particular that brought that pioneer seeker of agricultural origins, Robert Braidwood, to the eastern stretch of the Fertile Crescent. He searched the Zagros foothills for pioneer farming sites immediately below the places where mouflons and bezoar roam today. At the 9,000-year-old settlement at Jarmo he was rewarded, and not just by bones of the world's earliest domesticated goats alongside some very early cereals. The site also yielded numerous clay models of this animal that was clearly so important to this ancient community. While Braidwood's animal evidence was graphically enriching the story of a localized origin, animal bones were also to prove pivotal in the arguments favouring a more dispersed, evolutionary path to agriculture.

In Baluchistan, 1,500 km further east from the easternmost part of the Fertile Crescent in the Zagros mountains, beyond the Iranian Plain, the collapsed remains of ancient red mud brick dwellings spread out over twelve hectares of a hillside, perched up above a valley leading down into the Indus Basin. Those mud bricks were laid over 8,000 years ago, and beneath them are the remains of settlements stretching back a further millennium at least. Excavations at this site of Mehrgarh revealed huts and granaries, as well as the remains of some very early barley and wheat. There were also many animal bones, which have been subjected to intensive study by Harvard archaeologist Richard Meadow. In the site's early levels he found the bones of many of the wild forms of many domestic species, including sheep, goat, pig, horse and cow. In higher, more recent levels, those same species appeared in domesticated form. It seemed that the realm of pioneer domestication might have extended far beyond the Fertile Crescent to the south and east. In parts of the world even further from the Fertile Crescent, the geographical match between domesticates and their wild progenitors was still less clear.

The animal evidence was central to the arguments of Eric Higgs's group at Cambridge against a revolution that was restricted in space and time. They argued that the beginnings of animal management were not neatly divided into discrete episodes within centres of origin. Those beginnings were instead widely distributed. They were still proceeding, and had been proceeding well before the visible transformation of wild cereals and legumes in the Fertile Crescent. The group toyed with the idea that, even in previous interglacials, human management of reindeers and other herbivores was on a par with more recent management of domestic animals. Whether or not that speculative idea is persuasive, there is little doubt that the earliest instance of an animal modified by domestication has little to do either with the Fertile Crescent or with agriculture.

an ancient friend

The shorter snout and modified teeth that distinguish the domestic dog from the wild wolf have been identified in pre-agricultural sites as far afield as north-west Europe and Japan. Dogs and humans have to some extent domesticated each other, forging a hunting partnership advantageous to both. How ancient that partnership is we do not know, but we can try to find out by using the same evolutionary stopwatch on dogs as we have done on humans, the mitochondrial control region.

In fact, it is not exactly the same stopwatch. Because the control region evolves so fast, it varies a great deal between different types of mammal. Such variations are one of the main reasons why different species' mitochondrial DNA varies in length as well as in sequence. Nevertheless, the principle of using the control region as a measure of close evolutionary relationships is much the same in all these species.

At the University of Los Angeles, Carles Vilà embarked on a molecular survey similar to the one conducted on humans in the Mitochondrial Eve project. He took samples from sixty-seven breeds of modern dog, selected from all around the world. They ranged from Australian dingoes to Alaskan huskies, from Afghan hounds to English sheepdogs to New Guinea singing dogs. He assembled a parallel range of wolf samples, from localities scattered throughout the world. To root his phylogenetic tree, he also included coyotes and jackals in the survey.

The data from these wolves and dogs generated a family tree with a broad spreading crown. Across the breadth of this crown, the dogs clustered into four lineages, each of them with a considerable degree of internal variation. Individual breeds of dog did not match on to particular haplotypes, and a single breed might include several haplotypes. One of the lineages was of particular interest, as it included only domestic dogs, and none of the wild wolves. This looked like a lineage the story of which was completely contained within the history of that most ancient of partnerships, the domestication of the dog. The haplotype diversity was the key.

So far as we know from fossil evidence, the wolf and coyote diverged about 1 million years ago. The sequence divergence between them, which these data placed at around 7.5 per cent, could therefore form the basis of a molecular clock. If we compare that figure with its equivalent measured from the single lineage composed only of domesticated dogs, then some sense of the antiquity of the partnership can be gained. That second measure of diversity is around 1 per cent. If 7.5 per cent is what you get after 1 million years, then 1 per cent is what you might expect after 135,000 years, but let us not be deluded into introducing too much precision. We are, after all, relying here on that uncertain timepiece, the molecular clock. Nevertheless, whatever the imprecision of that estimate of 135,000 years, we would be hard pushed indeed to compress it within the 10,000-year period conventionally associated with farming. Our first domestication is much more ancient than that, and, judging from the spread and diversity of haplotypes, could well have happened at various times and places.

Soon after the publication of this result, one of Vilà's students at Los Angeles, Jennifer Leonard, set about placing this tree within its archaeological context. She made use of a piece of recent geological history that has been central to the conception and design of a number of studies in molecular archaeology. The two great landmasses on which most humans live, the Old World and the New, have on several occasions been linked by dry land between Siberia and Alaska. This has happened during each cold phase of the last 2 million years, as a result of the fall in sea level that accompanied expansion of the ice-cap. There was a series of discrete episodes when land animals could freely migrate from one landmass to the other. For this to be possible, it had to be cold enough for the sea level to drop and a land bridge to form, but warm enough for such a migration to be feasible. By the time the most recent of these episodes was coming to an end, around 14,000 years ago, our own species had made the journey from the Old World to the New, and so had the wolf or dog, in some form.

The Native Americans whom Europeans first encountered commonly raised dogs, and some of their breeds can still be found today. There is much we do not know about such ancient American breeds as the Alaskan Malamute, and we cannot even be sure that their ancestry is truly, or fully, American. After all, the European horse was speedily assimilated by the High Plains Indian. It could easily be that

European dogs came over in ships and were also drawn into the existing gene pool. To find out more about Native American dogs and their ancestry, Jennifer Leonard sought out dog bones that had been excavated from archaeological deposits that pre-dated Columbus. She managed to track down seven different pre-Columbian dogs from Bolivia and Mexico, and to amplify their DNA. They were quite diverse; the seven dogs generated six haplotypes which could then be positioned on the evolutionary bush that Carles Vilà had published a few years earlier. They gave a clear and significant result.

The pre-Columbian dogs all clustered together with their domesticated Old World relatives, and evidently shared a common ancestry. What this means is that the ancestors of these particular dogs did not make the long journey from Asia to America as wolves, but came with their owners as domesticated dogs. Moreover, they did so several thousand years before the term 'domestication' could be applied to any other species, in any part of the world. It is estimated that 14,000 years ago is about the latest that it could have occurred. Many believe that the journey happened much earlier, and certainly the journey from the heart of Asia was much earlier. Vilà's first estimate of over 100,000 years may well remain a reasonable time frame for the most ancient domestication of all.

The dog is a very special animal, but not alone in being treated with a particular kind of respect. In different parts of the world, other species – for example the cat, the horse and the cow – have been similarly accorded a special status in the human world. None the less, domestication of animals has less commonly been about companionship and more commonly about changing our relationships with the animals we kill and eat. Many of our victims have been the larger mammalian herbivores that our ancestors tracked and preyed upon for tens of thousands of years before domesticating and containing them. This is a group that would have had a rocky ride through the intense climatic fluctuations of the Quaternary Epoch, even without the added peril of human predation. Rapid climatic and environmental change generally favours smaller, faster growing things. It was often human predation that tipped the balance of an already fragile relationship between an animal and its changing environment. That is perhaps how the dramatic loss of indigenous large mammals from the New

World is to be explained. In other instances, the human predator turned farmer or pastoralist has saved large mammals from extinction by enslaving them. Each fated animal has its own story that can be tracked by molecular methods that reveal their histories and contribute to our understanding of that very human process of domestication. First, let us look at some of those that neither sought nor found refuge in human enslavement, but perished instead.

the perils of being large

Over the last 2 million years, global temperature fluctuations gained a fresh intensity, intermittently blanketing large swathes of the planet with ice. The period has been peppered with extinctions of mammals in the larger size range. They seem to have been especially vulnerable to the periods of global ice melt, when temperature changes were particularly pronounced. The most recent period of warming saw a swathe of such extinctions. Occasionally, domestication by humans provided a lifeline to these species. Some of our largest domesticated animals, such as cows, horses and camels, have, to the best of our knowledge, lost their ancestral relatives from the wild. Sometimes we find a wild relative in remarkable condition – a frozen wild horse, preserved for 30,000 years in the Siberian ice, or a 9,000-year-old wild camelid, desiccated in the Peruvian Andes. In each case, the hair and flesh is preserved along with the skeleton, and retains its natural colour. We can also find preserved remains of the species that did not make it, and found neither a natural refuge nor the protection of domesticated subservience to humans. One of these is a familiar feature of the Quaternary muds and gravels, which frequently yield its massive teeth and sometimes its enormous tusks. From time to time from beneath the Siberian ice and permafrost an entire woolly mammoth comes to light, with flesh and hair intact.

woolly mammoths, stopwatches, clocks and calendars

A number of extinct animals can now also be reached through their DNA. In previous chapters we have heard about the South African quagga, and our close relative the Neanderthal. Ancient DNA studies have also been conducted on the marsupial wolf from Tasmania; the giant sloth that once roamed in America; the dodo that travellers to Mauritius so rapidly exterminated; that gigantic flightless bird from New Zealand called the moa; an island tortoise; and the wild cow, which is featured later on in this chapter. Some of these, like the quagga and wild cow, are sufficiently close to living relatives that we can use the now familiar mitochondrial control region to judge the closeness. With others, it may be far less clear. The ancient giant sloth is clearly related in some way to its diminutive living relatives, as is the woolly mammoth to its naked elephant cousins. How close is uncertain. If they are very near, the control region will measure the distance well. If they are too distant, then it may get a little confused, with reverse mutations and deletions piling up upon and overprinting one another. A more slowly evolving stretch of DNA might give a better measure. In these terms, the control region provides a stopwatch, when what is needed is a slower-paced clock.

If we look on either side of the mitochondrial control region, stretches of DNA sequence are found that evolve at a slower rate. Immediately adjacent to the control region is a gene involved in energy management called the cytochrome b gene. Variations within this gene provide a suitable clock to compare animals that are broadly in the same genus or family. On either side of the control region are two genes involved in assembling the cell's RNA toolkit. They are referred to as the 12S rRNA and 16S rRNA. The numbers 12 and 16 are measures of the size of the RNA molecules they are programmed to build. The 12S rRNA gene provides another clock, measuring intermediate evolutionary rates, while the 16S rRNA evolves at a slower pace still, providing a calendar rather than a clock. Such an array of timepieces can be brought together to assess evolutionary relationships on many scales, and can be assessed together if the scale is unclear.

Erika Hagelberg settled for the cytochrome b clock when she decided to establish the presence of ancient DNA in these frozen mammoths. She took samples from two of the beasts that had come to light in the 1970s. One, from the Taymyr Peninsular, the northernmost extension of the Siberian mainland, was carbon dated to at least 40,000–50,000 years old. The other, from further to the east and 16m down into the permafrost around the Allaikha River, could have been over 150,000 years old, judging from associated fossil evidence. From both mammoths, she managed to amplify sequences from the cytochrome b gene of over 300 base-pairs in length. They may well be the oldest confirmed specimens to yield sound, amplifiable DNA.

The sequences were fairly close to those of living African and Asian elephants, and Hagelberg concluded that they formed a fairly tight group in evolutionary terms. Svante Pääbo was also attracted to the mammoth problem, but decided to work on the 16S rRNA gene calendar. He extended the sample to include five distinct mammoths. Four of these gave successful amplifications, allowing comparison with a range of different ungulates, including the elephants. The mammoths did indeed cluster with the elephants, but what was really striking about them was how much they varied. The two living genera, the Indian and the African elephant, differed in only two positions along the 93-base-pair sequence studied.This gene is after all a calendar, not a stopwatch. However, even in the small sample of mammoths studied by Pääbo's Munich team, there were differences along this stretch in up to five positions along the sequence. A tiny glimpse was caught of the great genetic diversity of these extinct mammoths, apparently dwarfing the diversity of living elephants. Looking just at the whole fossils, the impression gained was that two out of three elephants survived, all but the woolly one. On the basis of DNA evidence Svante Pääbo and his colleague Matthias Höss were emphasizing instead the diversity of ancient mammoths, of which all but a few – just a couple of naked ones – perished.

Shortly after this work was published, the picture was further clarified by bringing another one of these large, now extinct, mammals into the picture. This was the mastodon, another hairy relative of the modern elephant that was found throughout the world and probably preyed upon by the first Americans. Edward Golenberg, whose ancient

Magnolia leaf had been given a rough ride by Pääbo and others, now used mastodon evidence to take a critical look at Pääbo's own results with mammoths. Golenberg had sampled a well-preserved mastodon skeleton from a Michigan bog, along with a few extra mammoth fossils. What he and his colleagues were able to show was that, taken alone, the ancient mammoth and modern elephant DNA could generate a variety of virtual evolutionary bushes, according to slight fine-tuning of the computer program employed. It lacked robustness on its own. When another dimension of data was incorporated in the form of mastodon DNA, however, that robustness was gained. Those virtual bushes, when properly rooted in this way, resolved themselves into a single pattern. This pattern had the mastodon out on a limb as the distant cousin, and the mammoths clustering closer to the Asian than to the African elephant. This did not necessarily mean that Höss and Pääbo were wrong about the diversity of extinct mammoths, but that there was more pattern that could be gleaned by bringing the DNA of other extinct species into the picture.

Beyond the mastodons and the mammoths in the ice, a range of other large mammals struggled against the uncertainties of climatic change. One that can also be recovered from the Siberian ice has survived within the human sphere, although its ancestors are gone from the wild.

horse-riders of the steppes?

The frozen wild horses of Siberia, whose hair and flesh has remained intact for 30,000 years or more, are the most recent in a much older sequence of horse fossils, mostly surviving as bones, and going back tens of millions of years. These are the fossils that, in the late nineteenth and early twentieth centuries, provided one of the best sequences to support Charles Darwin's idea of gradual evolution by natural selection. The earliest members in this sequence, going back some 55 million years, are found in North America. By 2 million years ago, a fairly modern looking wild horse roamed the plains of North and South America as the fluctuating climate opened up vast stretches of grassland. Moving forward another million years, some of these horses made the journey across the Bering Straits, when the sea level was suitably low, to the grassy expanses of Siberia, eventually giving rise to our 30,000-year-old specimen. From Siberia they spread to Mongolia and Asia, arriving in Europe 750,000 years ago.

There are no wild descendants of our frozen horse. Truly wild horses disappeared from their ancestral homeland in America between 10,000 and 8,000 years ago. In Asia, they gradually dwindled in number, coming close to extinction in the last century. Those that entered domestication flourished and diversified, giving mobility to humans to a degree unmatched until the invention of the combustion engine. Overland travel on foot through unmanaged vegetation was always extremely slow. Although many prehistoric communities would have covered large distances during their lives, the rate at which they did so would have borne no comparison with that of the earliest horse-riders who traversed vast expanses of the later prehistoric world. We can get a sense of the scale of the transformation by following their sixteenth-century reintroduction into the New World. This fresh arrival of horses heralded an entirely new cultural tradition among the Plains Indians, and an entirely new set of ecological relations with the prairie and the buffalo. The great mobility of horse-riders has underpinned a number of hypotheses about some of the 'great journeys' of prehistory, to which we shall turn in the next chapter. Some investigators have followed Maria Gimbutas, in holding the horse, or more precisely its riders, responsible for the spread of Indo-European languages. She saw the domestication of the horse, in the wooded steppe north of the Black and Caspian Seas, as both the cause of, and the means for, a series of westward waves of horse-riding warrior societies. These waves took them across Europe, where they overran the more peaceful and harmonious matriarchal societies of Neolithic Europe. Prehistoric settlements such as Dereivka in Russia and Botai in Kazakhstan were spilling over with horse bones. From here her three waves of 'kurgan' (mound-building) cultures moved westward, spreading a new language, culture, technology and masculine ideology across western Eurasia.

This account of the radiation of the horse from an Asian centre of origin, taking with it the cultural package of its riders, is open to question on various fronts. Modern breeds seem to be very diverse, something that has led others to speculate about two, perhaps three separate domestications. These would separately account for breeds described as 'light' or 'warm-blooded', and those described as 'heavy' or 'cold-blooded'. The first group includes the rather elegant, fast and spirited animals such as the Arabian so favoured by horse-racers. The second includes the massive hard workers such as the Shire horse and the Suffolk Punch. Between these lie many crosses and intermediates, and there are the so-called 'native ponies' on the moors of south-west England, for example. According to one argument, the light breeds derive from wild horses from the heart of Asia, the heavy breeds arising independently from a European wild forest horse. Another argument gives native ponies their own local ancestry. This was clearly a problem to be tackled by genetics and ancient molecular evidence.

The palaeontologist Adrian Lister was keen to crack the problem and teamed up with geneticist Helen Stanley to do so. They collected ancient DNA from a variety of sources. Their most ancient material was the 30,000-year-old horse frozen in the Siberian ice, and bones of a similar age from Kazakhstan. Wild horse bones were collected from where they had lain buried for 12,000–14,000 years, in a cave in southern England, and in the tar pits of California. Others were selected from the 5,000-year-old site of Botai in Kazakhstan. The wild horse had lingered far longer than the woolly mammoth, and some available specimens were a great deal younger. The youngest 'ancient' samples were less than a century old, and came from the most recent horse breed to disappear from the wild. This was the 'tarpan', a small dun-coloured animal with a flowing mane and tail, which had survived through the Middle Ages in small herds in remote parts of central and eastern Europe. Some say there were two types of tarpan, a forest and a steppe form, but as the last known individual died in 1918, this is open to speculation. Fragments of its teeth and skin survived for Lister's and Stanley's analyses.

On the margins between ancient and modern DNA, they also had blood samples from a Mongolian wild horse hovering on the very edge of extinction. The Russian traveller, Nikolai Przewalski, had encountered this stocky animal, with its distinctive dorsal stripe, when travelling in the late 1870s through western Mongolia in search of undiscovered plants and animals. Thirty years on, twelve of these animals were captured and transferred to zoos, which is just as well, as the most recent expeditions in the area have failed to rediscover Przewalski's horse in the wild. The descendants of those twelve animals have provided blood samples to add to Lister's and Stanley's specimens. In addition to these various ancient specimens, they took samples from a range of living light and heavy breeds, as well as from a series of native ponies.

They studied both the mitochondrial stopwatch (the control region) and the corresponding calendar (the 16S rRNA gene) to get a sense of variation on different scales. From the stopwatch, they built a series of trees and networks that displayed a now-familiar pattern of deep-seated diversity. Working from a horse–zebra split between 2–4 million years ago, in this case based on a rich and detailed fossil record, the various living horse breeds were seen to stem from a common maternal ancestor that lived in the region of 880,000 years ago. Here was yet another phylogeny too deep to be contained within the archaeologically attested time scale for domestication. Those early central Asian sites, yielding the earliest archaeological evidence for intensive horse management, go back 5,000 years at most. There is no way that the sequence divergence measured by Lister and Stanley could be contained within this abbreviated time scale. Was this a split between light and heavy breeds? Apparently not. Two heavy breeds, the Suffolk Punch and Shire, are on completely different branches, and such breeds as the thoroughbred and Shetland pony are scattered among different branches. The Przewalski's horses do indeed cluster together, but not in a way that might suggest a long and separate history.

Lister's and Stanley's key finding was that the date for the common maternal ancestry of all living horses, including Przewalski's horse, corresponds well with the fossil record date for the first entry of the horse into Eurasia. Following this entry, and the genetic bottleneck associated with it, a huge range soon developed, and over the millennia a great deal of variation built up in their mitochondrial DNA, which is rather randomly distributed among modern horse breeds. Once again, the evidence for a major animal species in the human orbit is inconsistent with a very localized single domestication event.

Going back to those early sites in central Asia, we can see that diffuse and gradual process going on within particular culture groups.

When such sites as Dereivka and Botai first came to light, there was a tendency to seek within them the 'big event', the revolutionary transition from hunting wild animals to husbanding domestic livestock. A range of markers, from bone structure and wear on teeth to the simple quantity of horse bones, was used to argue the case. Marsha Levine has recently looked more carefully and closely at these sites, the population structure of the horses, and the evidence of stress and disease in the bones. She has arrived at a more complex story. The prehistoric inhabitants of these sites had a very long-standing and intimate knowledge of their horses. They understood their behaviour and movements, utilized them in a variety of ways and, judging from their elegantly carved horse bones, held them in veneration. But there is no particular evidence that the horses were other than wild in the conventional sense, and the point at which they were fully contained and bred within the domestic sphere is still open to debate. As late as the Iron Age, when horses were clearly being harnessed, ridden and widely used, the population structure of southern British horses suggests that many were still rounded up from a feral state for training and use. Against a behavioural background of this kind, the rather dispersed picture of domestication coming from the evidence of mitochondrial DNA need not surprise us.

crossing mountain and desert

More mysterious than either the mammoth or the horse is the camel. The two kinds of Old World camels have been fundamental to the human exploitation of the world's regions of greatest aridity and poorest vegetation. The two-humped Bactrian camel can be seen carrying loads throughout the highlands of central Asia. The one-humped dromedary also supplies meat, milk and wool from India across to the Sahara. Yet neither has left a trace of its ancestry in the present or the past. Nikolai Przewalski was convinced that he had found a wild camel as well as a wild horse, and one or two feral examples exist. None however has convinced the zoologists of its being truly wild and ancestral. To find any wild relatives, we have to trace their presumed journey out of Asia and back to the continent where the camelid line evolved.

Like the horse, the ancestral camelids crossed the Bering straits from their American homeland around 1 million years ago. The camelids that remained in America did not completely die out, and two species live on. These are the guanaco and the vicuña. As in the Old World, there are also two domestic camelids in the New World, the alpaca and the llama. What is more, the archaeological record is much more generous here.

In the arid desert that rises up above the Pacific coast, the same special conditions that have left us some of the world's best conserved mummified humans have also preserved the animals on which they depended. El Yaral was one of the sites from which the bodies of sacrificial llamas and alpacas were unearthed. Natural desiccation had left their skin and wool intact. Sometimes, their skin hung around the bone like a loose overcoat, but the muscles and internal organs were also frequently conserved. These animals were sacrificed by a blow to the head 1,000 years ago, and buried beneath the floors of the houses at El Yaral. A millennium later, their wool has lost neither its colour nor its texture. It is preserved so well that it can be analysed today using the same techniques as are applied to modern llama and alpaca fibres. The ancient wool was fine, finer than most produced today. From studies of these ancient fleeces, we sense that the farmers who dwelt here 1,000 years ago herded animals from highly selected breeds, breeds that no longer survive. Here was evidence that history, in particular the repression by the European conquistadors, had taken its toll. Its impact on indigenous human communities had also led to a loss of genetic diversity among their domestic animals.

At the research division of the London Zoo, Helen Stanley and Miranda Kadwell embarked on an exploration of the DNA within these specimens. They speculated that perhaps the ancient DNA in these ancient llamas would reveal that genetic diversity. If they could get at the DNA in this well-preserved specimen, perhaps it might even be possible to recover the lost genes. Unfortunately, their only successful amplification of the ancient camelids was a sequence of fewer than 100 base-pairs from the 16S rRNA calendar. It was enough to show that the ancient llamas were closer to alpacas than to Old World camels, but the missing clock and stopwatch were really needed to take the study further. One of the points reinforced by the evidence from these ancient Peruvian animals was that tissue that seems visually intact might not be the best preserver of large molecules. In the right conditions, an ordinary bone may have far better preserved DNA than one on which the colour of an animal's hair is still visible. Even in these wonderfully preserved specimens, this remarkable molecule from the past remained elusive.

Each of the above accounts is of humans and animals interacting in parts of the world outside the Fertile Crescent. In many ways, the key molecular studies for this ancient region are of the sheep and the goats that first drew Robert Braidwood to the Fertile Crescent, but their story has yet to be told. However, a third animal from the region has revealed the most detailed molecular story of all.

the mighty urus

In the late fifteenth century, one Johann Pruess compiled the following description of a rather terrifying wild beast:

> Isidor says of the urus: urus are wild cattle so strong that they can lift trees as well as armed knights with their horns. They are called urus from the Greek word oros meaning mountain . . . Heylandus says . . . In the Hercynian forest of Germany the Urus is found. These animals are nearly as large as elephants: in appearance, colour and conformation they are like cattle. The force of their horns is great and their speed is great. They spare neither man nor animal. One catches them in pits and kills them.
>
> (Johann Pruess, *Hortus Sanitatus*, 1495)

None of these mountainous beasts, now referred to as 'aurochsen', survives today. They live on only as descriptions, illustrations, fossil footprints across an ancient beach, and in archaeological deposits, their gigantic bones and particularly their massive horns reminding us of Pruess's words.

Something of the feelings they inspired in prehistory can be imagined at an ancient site in the Konya Plain of southern Turkey. Discovered by James Mellaart in the 1950s, Çatal Hüyük remains one of the earliest sites to show evidence of cattle husbandry. As Mellaart dug down into the vast thirty-two-acre tell he uncovered a crowded cluster of mud-brick houses. Within the sediments was a diversity of artefacts unparalleled in a settlement that was just a few centuries younger than the earliest permanent farming settlements in the region. Among the finds were the massive spreading horns of aurochsen that had clearly inspired the inhabitants with great awe. In one of the rooms that Mellaart suggested was a shrine, the space was dominated by models in clay of bulls' heads around the walls, painted in striking colours and fitted with genuine horns.

Çatal Hüyük is now being re-excavated by Ian Hodder, and a team at Trinity College, Dublin, is currently attempting to amplify the DNA from the cattle bones excavated from the site. At the time of writing, the search for these particular DNA molecules has not reached completion, but Dan Bradley's team already have more than enough molecular evidence to question the single origin of domesticated cattle. Just as Richard Meadow had looked beyond the Fertile Crescent, further to the east and into the Baluchistani mountains, the Dublin team also looked eastward.

Going beyond Meadow's site of Mehrgarh and into the Indus Valley itself, we find the impressive urban sites of the more recent Harappan civilization. Harappa, Mohenjo-Daro and the recently uncovered Dholavira were large towns of the third millennium BC, with streets, public places and impressive systems of water management. We still cannot decipher their writing, but from the images on their cylinder seals we can see how very important cattle had become in their lives. But these cattle are not quite the same as the ones found further to the north and west. Two particular features distinguished them from cattle found throughout Europe and parts of northern Asia. The Harappan illustrations displayed a prominent hump above the neck, and the pendulous flap, or dewlap, below the throat. The same features can be seen in living breeds, particularly in Africa, and in south Asia where they are known as zebu cattle. The difference between zebu cattle and their 'taurine' counterparts has been much debated by cattle breeders. The two forms are totally inter-fertile, and we have already seen in both plants and animals that even greater differences in physical appearance than this need not imply a great genetic distance. However, the molecular research conducted by Dan Bradley and his colleagues indicated that, in this case, an easily observed physical feature of the beast was a very good indicator of what was going on at the molecular level.

They worked with the mitochondrial stopwatch, the control region, and targeted a hypervariable segment of 375 base-pairs within it. In that stretch, thirteen different breeds of cattle displayed variation at sixty-three positions. When the resulting variations were built into a tree, it was a tree with a very deep bifurcation. On one major branch were the Indian humped cattle, and on the other all the remaining breeds. It was a clear division and an ancient one. After including bison in order to root the tree, they estimated that the divergence between humped Indian cattle and the rest was three-quarters of the divergence between cattle and bison. That split is assumed from fossil evidence to have happened over 1 million years ago. On the basis of modern DNA alone, there was a strong argument for separate domestications of cattle in south Asia and at some other location. The African cattle presented a bit of a puzzle.

Some African cattle have humps and others do not, but, with or without, they all fall on the 'un-humped' side of the mitochondrial tree. The distinction has a pattern both in space and in time. The humped breeds are a more southerly phenomenon, the northern breeds lacking the hump. Back in time, so far as we can tell from historical records, this distinction was even more pronounced than it is today. In prehistory, things were different. Domestic cattle were in Africa at least from the seventh millennium BC, and at that stage all lacked the characteristic hump. Before the Sahara dried up, the cattle that grazed on its natural pastures were captured as images in the Tasili rock art, images also free of humps. It is not until around 2,000 years ago that the distinctive vertebrae of a humped cow had found their way into the Neolithic site of Ngamuraik in Kenya. Around three centuries later, an Axumite figure of a humped cow is found in Ethiopia.

These times correspond to a period of considerable change for the southern regions of the African continent. Where all the known evidence of human activity had indicated hunting and gathering, pottery, iron-working and farming all spread southwards across regions now occupied by Bantu-speaking peoples. This expansion and spread of farmers/metalworkers is held to have transformed the subsequent history of southern Africa. While evidence for cattle remains scanty after those first finds of humped cattle bones, they may have had a key role to play in the southerly spread of later generations of pastoralists. Humped cattle have a metabolic advantage over their unhumped relatives. They can slow their metabolism more effectively, something that may be key to survival in a savannah environment. Indeed, the adaptive strength of the savannah grasses themselves is that they can drop metabolism to zero by drying up and turning to seed. Humped cattle and savannah grasses alike are well adapted to the fluctuating environments across which those early pastoralists travelled in the south. What the archaeological evidence indicates is that there were humpless cattle in the more northerly regions, and two groups of humped cattle – one in south Asia and one in southern Africa – with a separate but uncertain past. Bradley's group built up a larger collection of African specimens but still came up with the same result. All African cattle, regardless of their appearance, with or without humps, fell on the other side of the tree from the Indian humped cattle. So where did the genes for the hump come from?

What the mitochondria tell us is that the maternal lines of African cattle all lead back to a humpless, taurine form. The hump must have been inherited from the male line alone, from zebu bulls. Indeed, that is the only way to account for the evidence. At some stage in African prehistory, certainly by 2,000 years ago, some Asian bulls had been brought across to this continent. They looked very different, but someone must have known that the special hardiness of the Asian cattle was something that the African cattle lacked. For a long time, those breeds had a fairly localized impact, but by the time the spread of pastoralism was transforming the southern continent, the hardy genes they carried were absolutely key. This much is conjecture, based on the absence of zebu mitochondrial DNA. The proof lies in the chromosomes only transmitted through the male line.

Bradley's group went on to study the paternally inherited Y-chromosome in European, Asian and African cattle. Their hypothesis was confirmed. In Europe the animals were consistently taurine, in south Asia they were consistently zebu, but in Africa the Y-chromosomes were a complex mix of taurine and zebu. What seems to have happened is that zebu cattle were indeed brought to central Africa some time before 2,000 years ago. It is difficult to say how many were brought, or whether it was bulls and cows together, or bulls alone. It is the bulls that left their mark by interbreeding with the local cows. Their hybrid offspring then travelled and mingled across central and southern Africa, with the Bantu-speaking peoples of the Chifunbaze culture, whose sites appear from the late first millennium BC and proliferate through the subsequent millennium. These cattle began as a small component in their economy, but eventually grew into a major feature of it. The appearance of the zebu cattle in Africa is complementary to the appearance in third and second millennium BC India of African grain crops such as sorghum, which in turn transformed the nature of Asian agriculture. It used to be thought that such intercontinental contacts ran through the great civilizations of the Indus Valley and the Near East. These cultures may not however have had much to do with it. The relevant bio-data does not follow the inland route through the Near East, but links sub-Saharan Africa with mainland India via a few points on the intervening coast. It must have been a less celebrated maritime community that took crops and livestock back and forth, a few exchanges transforming the human ecology of both continents. Moreover, the most recent genetic results from Dublin suggest that the geographical heart of the zebu haplotype is in southern India, rather than towards the Indus valley in the north.

By the time a Chifumbaze settlement just outside Fort Victoria had grown to the monumental, high-walled Great Zimbabwe after which its country was named, this trade with the east had become extensive. The site contains Near Eastern glass and pottery from as far away as Persia and China. But the most critical import from the east was not those conspicuous high status goods, but a few genes that had arrived a few centuries before the massive stone buildings were erected. A group of zebu cattle, perhaps exchanged for sacks of African grain, had transformed the great herds of cattle upon which that society depended and the human ecology of a subcontinent. Furthermore, we only know this from the evidence of the molecular tachometer encapsulated within their cells.

One final part of the story is the history of the taurine branch of the African cattle breeds. Their conventional archaeology is in no way inconsistent with a spread from the Fertile Crescent, down the Nile

Valley and across land to north and west Africa. However, the growing understanding of D-loop data allowed Bradley's group to go one stage further and look for the common ancestor of Asian, European and African cattle. What they found was that the three continents contained three haplotype clusters. As we have seen, the common ancestor of European and Asian cattle lived between 200,000 and 1 million years ago. They estimated that the equivalent common ancestor of European and African cattle lived 22,000–26,000 years ago – much closer related than taurine and zebu, but still too far back to accommodate a common single ancestor within the last 10,000 years.

By looking further into the patterning of the D-loop variations, it was possible to tease out the demographic history of these variations. Wild cows spread out from the refuges they had occupied during the coldest period of the most recent Quaternary cycle, into Europe and Africa, leading to genetically divergent populations of wild progenitors. The wild zebu had diverged during some earlier Quaternary cycle. In all three continents, cattle were independently domesticated. One group of domesticates spread westward across north Africa from 9,000–11,000 years ago. Another group spread across Europe 5,000–9,000 years ago, and a cross between African taurines and a population of imported zebu bulls spread across sub-Saharan Africa over the last 2,000 years.

The genetic precision of this story is impressive, but we must remember that the chronological precision is still no better than the molecular clock on which it is based. To reiterate: the basis of the clock used in this case is based on two assumptions. First, it is assumed that *Bos* (the cattle genus) and bison diverged 1 million years ago, and second, that the mutation within the sequence of 370 base-pairs studied is relatively uniform. It works out in fact at one base-pair mutation, on average, every 4,000 years. Even if both these assumptions are sound, the difference between African and European cattle is only two base-pair mutations away from a single domestication. Once the molecular clock is measuring differences of hundreds of thousands of years, then it is on firmer ground. Once the differences of interest reduce to a few thousand years, then what are needed are more precise dates from archaeology, and DNA sequences from ancient bones, both domesticated and wild.

The bones of domesticated cattle are widespread, and from time to time the bones of the ancestral wild aurochsen turn up, for example in Upper Palaeolithic cave sites. The Oxford group of ancient DNA researchers tracked down a number of these aurochs skulls, and attempted to extract and amplify aurochs DNA. Four samples from caves in southern England, ranging in age between 11,000 and 12,300 years, were assayed. The European aurochsen were far closer to the modern European taurine cow than either were to the south Asian zebu, a result subsequently corroborated by DNA from a number of other aurochsen. The genetics of these extinct beasts was adding clear support to the multiple domestication hypothesis.

These data from modern and ancient cattle, domesticated and wild, allowed another tree to be built, rich in information about the cattle's past. A principal feature of the tree was the three distinct clusters, corresponding to distinct domestications in south Asia, the Near East and north Africa. The internal diversity within each cluster indicated several millennia of independent divergence in each. Then the trajectories of each could be independently followed into different regions of the Old World.

One region of particular interest was Europe. Here the taurine cow from the Near East could be traced, with a strong cluster including such breeds as Charolais, Simmental and Friesian bearing very similar mitochondrial sequences. Outside this cluster, another group of European cattle was out on a limb, and a limb that retained a surprisingly high diversity. These were the breeds from the north-western islands of Europe, including British and Channel Island breeds. Their mitochondrial divergence suggested a considerable antiquity, which puzzled Bradley's team. One possible explanation is that some of the domesticated bulls on these islands had crossbred with indigenous island aurochsen, from which their divergent mitochondrial haplotypes had arisen. There is no problem with this genetically, now we know how very closely related the two were, but there is more to coupling than DNA. Rereading Isidor's and Heylandus's observations of the strength, size, speed and massive horns of the aurochs brings to mind how brave that diminutive ancient proto-Friesian bull would have needed to be to seek out such a mother for his calves. The idea of crossing with aurochsen was further questioned by the sequences now coming from a range of aurochs skeletons. Their mitochondria did not match up with those of the island breeds. It was more likely that the spread of domesticated cattle came in several stages, with successive waves sometimes – but not always – over-stamping earlier genetic patterns.

These animal stories add another dimension to our understanding of farming beginnings. Plant and animal domestication have often been treated as part of a single process, an idea reinforced by the coexistence of so many precursors of major world crops and livestock in a single region, the Fertile Crescent. The DNA evidence separates them a little more. In both cases, the pattern of domestication is patchy in space and time, but with animals the situation is more dispersed still. Virtually every animal domesticate so far examined through DNA methods would appear to be the consequence of multiple domestication highly dispersed in space and possibly time. The domestications of dog, cow and horse were processes scattered across entire continents, a far cry from the single-centre accounts.

Something else can be said about these last three animals. Their domestication did not just influence humans through the food in their stomachs, but also changed their mobility. Dog, cow and horse each allowed humans to travel vast distances together with sizeable loads. They could pull a sledge or cart, and the larger ones could pull or carry people as well. The control of animals as well as plants initiated a new set of journeys, leaving yet another series of molecular tracks in their path.

7

great journeys

leaving the cradle

The idea of great journeys blossomed in the nineteenth century and has dominated twentieth-century ideas about the human past. At one point, the idea of 'progress' existed in the European mind largely detached from the constraints of geography. Progress transcended place. Human communities around the world were simply at different points on a common journey of progress, a journey that could be described as unilinear evolution, a path of improvement along a single identifiable route. If communities had not yet turned to agriculture, then they soon would do, perhaps with a little colonial help. Towns, trade, monuments and the trappings of civilization would naturally follow. During the twentieth century, we have been more circumspect about the inevitability of progress, and more aware of variation, conflict and political upheaval in human history. Rather than envisaging the steps of progress as universal benchmarks, we see them instead more as historical episodes, transitions such as the beginnings of agriculture occurring at a particular place and time. Following those historical episodes are the journeys outwards from the point of origin, taking new ideas and new ways of sustaining life to far-flung regions, through contact, colonization or conflict.

Twentieth-century archaeology has continued to be absorbed in these journeys, not always the same journeys, but the underlying concept of movements and migrations, as ways of explaining changes in human culture and society. In a previous generation, there seemed nothing odd about that pioneer of archaeological fieldwork, General Augustus Pitt Rivers, charting the progress of humanity from excavations on his own Dorset estate. Why look further than your own backyard for a process unconstrained by geography? By the time the General was assembling his grandest typological schemes of linear evolution, however, the premise of inevitable progress was already coming to seem a little unrealistic, particularly among a younger and more cosmopolitan generation.

At the start of the twentieth century, at a moment when some quite astonishing discoveries about ancient Egypt were coming to light, a young Australian took the post of Professor of Anatomy at Cairo University. Grafton Elliott Smith decided to apply his scientific skills to the growing number of mummified bodies becoming known, while his fascination with the sophistication of their preserving methods and the wealth of their graves began to grow. In a land where wild barley sprang spontaneously from the Nile muds, he came to the view that the cemeteries of Egypt were the birthplace of the arts and crafts of civilization. He had the opportunity of returning to Australia in 1914, and of inspecting another group of mummified bodies from the Torres Straits Islands. What he saw convinced him of a link between the two groups. The methods of preservation used on these bodies from Australasia seemed to bear witness to the influence of the ancient Egyptian cultures on the far side of the world. In the following years, he traced patterns of similarity in a whole range of cultural features that pointed to contact. These included the practice of mummification, massive stone architecture, symbolic representation of the sun and the serpent, and the use of metals and domesticated plants and animals. Each of these features could be followed through the Old World and the New, and even to remote islands of the Pacific. The idea of a global diffusion of ideas was born, rooted in Egypt and reaching out to the ancient civilizations of the world.

Smith's grand scheme never gained universal acceptance. The old guard saw their unilinear schemes of progress under threat, and a younger generation was keen on emphasizing the contrasts rather than the similarities between ancient cultures. However, Smith had paved the way for more modest schemes, ones that would succeed in providing alternative frameworks to unilinear evolution. One came from his fellow Australian, Vere Gordon Childe, whose contribution to the debate on agricultural origins was discussed in Chapter 5. At one stage

Childe was quite interested by Smith's hypothesis, but in due course his own encyclopaedic knowledge of world prehistory would enable him to come up with a more modest, but also more persuasive model. In this, he traced the spread of agriculturally based, urban civilization from the Near East across Europe. Within a more regionally compact framework such as this, the idea of diffusion of culture became the predominant archaeological model. In the middle years of the twentieth century, charting the pathways taken by those ancient migrants, or the ideas they passed on, became the prime quest of the prehistorian. Arrows proliferated across maps, linking artefact or monument types along a common path. Not everyone traced their paths in the same direction or within the same time scale. Before the widespread use of scientific dating methods in the 1960s there was a fair amount of leeway, both in what evidence was used to trace the path and in what date was ascribed to it. In Germany, Gustav Kossinna had provided what would become the prehistoric prelude to the Nazi version of history, surmising an Aryan rather than an Oriental source point for the arrows across his map, for the great civilizing journeys. Several alternative journeys were put forward to account for the commonality of particular language families, most notably the Indo-European language group that spanned a vast stretch of Europe and Asia. As we saw in the previous chapter, Maria Gimbutas drew together evidence from artefacts and weaponry, burial traditions and language elements to reconstruct three waves of expansion by horse-riding warriors between the fifth and the third millennia BC, carrying Indo-European language and transforming European culture on the way.

The decisive test for many of these great prehistoric journeys came when the dates of ancient monuments and prehistoric civilizations around the world no longer floated free, to slip effortlessly into one grand narrative or another. From the 1960s, a battery of independent scientific dates came into play and forced the world's antiquities into authentic and verifiable time slots. Several of the arrows snapped in the process and had to be removed from the maps; the dates along their paths were all wrong. Armed with these new dates, the archaeologist Colin Renfrew took on the most established part of the story.

Europe as an archaeological study area has its pros and cons. Nowhere else has so much data been gathered about the human past.

The stories we can build around this vast body of information are steeped in detail and complexity. However, they are also steeped in established views held by the continent's unusually high density of senior academic figures, which may at times offer the young researcher a greater constraint than the data itself. What is more, Europe is the continent where the western notion of civilization was born, and people care about getting its myth of origins right. Not for the first or last time, redrafting the European story was no easy task. None the less, Renfrew's critical use of both the growing body of carbon dates and the cultural evidence dispatched some of the best-known diffusionary arrows from the map. By the early 1970s, a new approach to the origins of European civilization was required.

To set this reconstruction in motion, Renfrew brought together a range of specialists at a conference on the explanation of culture change at Sheffield University. Among the speakers was the eminent geneticist, Luca Cavalli-Sforza. The new archaeologists of the 1970s had a great enthusiasm for predictive models from science, and Cavalli-Sforza offered one. It was a mathematical model that looked at the behaviour of a population undergoing sustained growth. Even without any conscious aim of migration, such an expanding population would in any case project a 'wave of advance', merely as a consequence of a large constellation of small journeys away from the over-populated core. He teamed up with the archaeologist Albert Ammerman to see if this worked with radiocarbon dates, and it did. The dates for the early farming sites also fell into a sequence of diminishing antiquity, moving outwards across Europe from the early agricultural sites in south-west Asia.

Once again, a wave of farmers was on the move, but this time on a schedule that matched up with the rapidly growing corpus of radiocarbon dates. Ammerman and Cavalli-Sforza concluded that, once domestication had occurred in the Fertile Crescent, the ecological advantage of farming then led to population growth. From this growth at the core, there emanated a wave of advance across the continent, tracked by carbon dates and genetics alike. According to this new working of the data, the diffusionists had focused on the wrong type of movement and too late a cultural episode to match the dates. It was not monument-building, metalworking, farming communities that actively carried the wave front forward. It was their distant ancestors, the very first pioneer farmers, responding to the pressures on the land they left.

The existing diffusionary journeys had been dispatched, but a new set of journeys was on the horizon, based on fairly new scientific data that were still only sketchily understood. None the less, both data-sets were about to expand rapidly. The number of carbon dates was still increasing, and Cavalli-Sforza was persuaded the time was ripe to push ahead with collating the genetic data. In 1978 he embarked on an eight-year collation of molecular genetic data from living humans, at the end of which time he had persuaded Colin Renfrew and others that his genetically traced journeys were well worth taking seriously.

prehistory from gene maps

Before DNA science fully came into its own, some of the most reliable information about genetic groupings of humans came from blood groups. These groups relate to slight variations in the protein structure of certain components of blood, which in turn arise directly from genetic variation. They range from the well-known ABO groups, through variations in an important group of proteins known as the immunoglobulins, to some obscure variations little known beyond the medical world. By the time Cavalli-Sforza set to work compiling and collating information on human genetic diversity, there were data on numerous protein groups displaying genetic variation, for which medical and anthropological records allowed a global survey. With his colleagues, Paolo Menozzi and Alberto Piazza, he set to work gathering data for all the gene loci which had been studied and charted in sufficient detail for global comparisons to be made. Each of these coded for a particular blood protein or enzyme, which varied between different people.

An important aspect of such variation is that it doesn't disrupt bodily function. Three people in the same community could quite easily be of A, B and O groups respectively without knowing they were different in any way. This apparently non-functional nature of the variation renders it an ideal tracker of lineage, undeflected by ecology and selection. The geneticists gathered information on protein variations from about 500 different human populations around the world, and set about building trees and constructing maps to represent the global origins of human genetic diversity. What emerged was a masterpiece of genetic analysis, placed in the context of a meticulous survey of human archaeology and language history. It provided a synthesis of great journeys in the human past that matched the scale of Smith's thinking, but with one key advantage over that earlier author's argument. This synthesis was in keeping with the scientific evidence for the dating of the human past.

An example of what they achieved is provided by their new enhanced analysis of Europe. They now had information on almost 100 gene loci for the continent, all varying in slightly different ways. To present all this variation would require a diagram of around 100 dimensions. It simply could not be displayed on a two-dimensional map. Fortunately, a statistical procedure exists that can flatten such multidimensional monsters into a two-dimensional shape. It is called Principal Components Analysis (PCA), and the flattened pattern so produced is called the 'first principal component'. The sheer process of flattening not surprisingly squeezes out a great deal of the multidimensional information. Indeed, quite commonly more is squeezed out than is left in. The PCA can do something with that, however, by going for a 'second pressing' to produce a 'second principal component' that accounts for another portion of the information. The sequential pressing can go on until it is judged that little information remains to be squeezed out. What sounds like olive oil production is, of course, a number-crunching exercise proceeding within a computer. Like the series that ranges from a first cold pressing of 'extra virgin' oil through to a series of cruder, less expensive oils, PCA produces a series of principal components, each successive one accounting for a little less of the desired information. On a map of Europe, the first principal component of genetic variation forms a pattern resonating with Cavalli-Sforza's earlier result with Albert Ammerman. A series of bands runs from south-west Asia to north-west Europe. The idea of an expanding wave of farmers moving across Europe was holding up. The second principal component highlighted a community on the fringes of this wave, and not actually a part of it. The reindeer-herding

Saami from northern Scandinavia (formerly known as the Lapps) had followed a separate path from north-western Asia. The third principal component seemed to the authors to mirror another popular expansion, that of Maria Gimbutas' horse-riders of the steppes. Each principal component was elegantly pointing to a particular expansion, migration, or set of journeys.

Each of the other continents in the world was mapped in a similar way. Genetic variation was broken down into a series of maps displaying the first, second, third principal component and so on. Each of these twenty or so maps relayed a story of one or more movements of ancient peoples, carrying with them genes, language and culture across the land. Two major types of movement stand out in their interpretation of these maps. The first is the series of population movements that first brought anatomically modern humans to particular landmasses. The second is the series of population movements that brought pioneer farmers to take their ecological advantage to new niches, often replacing hunters and gatherers in their path. In terms of the human past, twentieth-century archaeologists ended the century as they had begun, by explaining the global diversity of human societies in terms of a simplified model of population movements and transfer of ideas. At the beginning the explanation was based on skulls and objects; at the end it was based on molecular genetics and radiocarbon dates. Today we can also reach back into the molecular genetics of the past and ascertain whether the ancient travellers themselves conform to that new global story.

first steps across america

One of these journeys had been a source of fascination long before Grafton Elliott Smith. The contact that followed Columbus's discovery of the New World opened European eyes to the fact that not only was there another landmass beyond the Old World, but somehow people had arrived there before Europeans had conquered ocean travel. Several centuries after Columbus set sail, we now know that America was not always encircled by water. In each cold phase of the last 2 million years, Siberia and Alaska were linked by a considerable tract of dry land, exposed by a drop in the level of the sea. Nevertheless, no hominid species apart from our own made the crossing, and it was a late crossing at that. The earliest dates for evidence of humans in the New World go back around 30,000 years, and even these are contested. Some would argue that the record is only half that old, at most. Unlike the other journeyers who feature in this chapter, the first people in the New World had not yet turned to agriculture, but we now know that they did have one domesticated species with them, the dog. That much is clear from the DNA evidence considered in the previous chapter. This most ancient of domestications may have played a key part in enabling those ancient people, more accustomed to the middle latitudes, to survive in the frozen, windswept landscapes of the northern crossing. Their dogs could pull heavy burdens on sleds as they moved across hazardous terrain, and would have been vital in helping humans cull the migrating herbivores that grazed upon the Arctic sward. Whenever the first crossing was made, it appears that by around 12,000 years ago, their descendants had reached America's southern tip. Now modern humans occupied all the world's major habitable landmasses.

America's first inhabitants did not leave an elaborate material trace behind them. It is not really possible to construct a detailed story of migration and colonization from their artefacts alone. Scholars have turned to other sorts of patterns, including those arising from language and genetics. What they have looked for is a set of wave fronts, something akin to the traces the tide leaves on a beach. If we carefully mapped the arcs of flotsam described upon the sand, then with a bit of luck the story of each successive tide could be charted. That is what Christy Turner thought when looking for variations in the teeth of living Native Americans. The teeth are clearly the most accessible part of the skeleton in living people, and are presumed to contain a range of genetic signatures. Turner arrived at a three-way classification that might correspond to 'arcs in the sand'. The arc closest to the northern crossing encompassed two northerly groups: the Inuit and the Aleut. The latter were the inhabitants of the final remnants of the submerged land bridge, the Aleutian islands that reach out from Alaska in the direction of Siberia. Further to the south, a second arc encompassed inhabitants of the north-west coast. The third and final arc drew in the rest of the New World.

Turner's pattern caught the attention of a historical linguist, Joseph Greenberg, who saw a similar pattern in Native American languages. He grouped those numerous languages into three 'super-families' that displayed a rough match with the dental groups of Turner. In the north were the Inuit and Aleut. In the north-west and in a few places further south, Greenberg recognized a group of 'Na-Dené' languages. The rest of North America, and all of South America, was rather contentiously lumped together to match Turner's third arc, and referred to as the 'Palaeo-Indian' group. At first, these two patterns shared a fairly uncertain relationship with genetics, but as the protein and DNA evidence was brought into consideration, some could see a match. Certain blood proteins mirrored the division, and the DNA displayed a north–south banding of some sort which, with a little bit of gentle persuasion, could be coaxed into the three-wave pattern. Even before that DNA evidence was available, Greenberg could speculate on the actual pioneer journeys into America.

One of these journeys began from one of the north-flowing rivers that drain from the Siberian mountains into the Arctic Sea. From the River Lena, this journey passed through the heart of ancient 'Beringia', the name given to the vast expanse of land exposed by the lowering of the sea. Occasional herds of large game, upon which the travellers preyed, crossed this region of arid Arctic steppe. Those travellers would be the ancestors of the 'Palaeo-Indians'.

Another journey began from the regions between these two great rivers, the Lena in the north and the Amur in the south. Here, foraging communities lived among the cold valley woodlands, very different from the distant Beringian steppe. However, as the climate warmed and ancient Beringia began to drown, the steppe gave way in many places to a mosaic of wooded and watery land, a new kind of corridor to which the valley woodland peoples were already acclimatized. These would be the ancestors of the Na-Dené speakers, who established themselves at various points in America's north-west.

A third journey started in the lower reaches of the Amur River, which separates modern-day Manchuria and Siberia. The fisher-foragers from this region made their way along the land ridge that separated the Sea of Okhotsk from the North Pacific, now a string of islands but once a substantial land corridor. From there, they continued to hug the south coast of ancient Beringia on their way to the New World, sustaining their fishing and foraging traditions throughout. These would be the forefathers of the Aleut and Inuit.

The new evidence coming from mitochondrial DNA then bolstered Greenberg's account. The global mitochondrial surveys had indicated that just a few branches of the human evolutionary bush had made their way into the New World. There had clearly been a genetic bottleneck, and it seemed reasonable to assume this related to the difficult passage across the frozen north, those pioneering journeys that Greenberg had described, across the land exposed during episodes of low sea level. This bottleneck allowed only a small cluster of lineages through from Asia. The geneticist Douglas Wallace identified four mitochondrial lineages, which he labelled A, B, C and D. They were distinguished by a small number of recognition sites around the mitochondrial genome, and in one case by the presence or absence of a 'deletion', a mutational loss of nine base-pairs at one point along its length. Within America, there was a clear trend towards a north–south banding of these haplotypes, with A most prominent in the north, B in the centre and south, and C and D becoming more common towards the south. All four lineages were found to be present within the most diverse group, the 'Palaeo-Indian' speakers, while the Na-Dené speakers were all of lineage A, and the Inuit and Aleut were either A or D. It was not a simple match, but it none the less seemed to offer support to Greenberg's three journeys. Now that DNA had been brought in, the molecular clock might also date those separate journeys. Wallace dated the first journey between 26,000 and 24,000 years ago, the second journey between 15,000 and 12,000 years ago, and the third journey at between 9,000 and 7,000 years ago. An engaging account, but not everyone was convinced. A number of counter-arguments were arising from a closer examination of DNA variation.

Among the critics of this account was one of Douglas Wallace's students who, like a good student does, went on to challenge his mentor's ideas. Andy Merriwether shifted his attention from the Native American communities in the front line of European contact, to a tribe which had managed to survive in the remote depths of South America's Amazon forest. There is a group of Indians living around the northernmost reaches of the Amazon River in Brazil known as the Yanomami.

Merriwether checked their mitochondrial lineages to see whether this remote community too could be brought into the four-lineage pattern. After sequencing a number of individuals, it became clear that each of the four lineages, A–D, was well represented. However, he was not convinced there were only four. It seemed there were seven such lineages, perhaps more. While such a remote people might be expected to harbour the occasional rare lineage, a closer inspection of other Native American groups was revealing other occasional deviations from the four-lineage account. Looking across the continental data, Merriwether arrived at the observation that all four lineages were widespread, but peppered across the map were instances of rarer lineages. It seemed to him that there was a much more straightforward explanation than Greenberg's and Wallace's three journeys. A single journey could account for the whole picture, the geographical variations being simply the consequence of sampling. Imagine a pack of fifty-two cards, each with the rather unusual ability to reproduce. Scatter those cards across the Bering Straits into America, and let them fall into clusters or 'hands'. One hand may contain rather a lot of hearts, another have no kings, and so on. If these hands reproduce mainly locally, then those founding biases will be retained, and that is how Merriwether argued that a single entry could produce the kind of variation recorded in the living population.

Meanwhile, a young molecular biologist working in the German University of Hamburg believed he could get closer to the issue by shifting the emphasis from the handful of DNA marker sites used in Wallace's argument to one of the most informative regions of the mitochondrial genome. Peter Forster turned his attention to the DNA sequence that has been key to so many of the projects discussed here, the first hypervariable segment of the mitochondrial control region.

Small variations within this fast evolving segment allowed Forster both to separate the founder lineages and to explore the details of their evolution. Models could be built of how they were related, and what had happened to them, before and after passing through the genetic bottleneck. These models were presented as networks of evolutionary pathways, and the networks so produced tallied better with some migrational stories than with others. Forster scrutinized his networks carefully, and concluded that all known founder lineages arrived as part of the same episode. He employed the molecular clock to place that episode around 20,000–25,000 years ago, spreading to the tip of South America by 13,000 years ago.

While this core network accounted for all founder lineages, it appeared to be overlaid by a second network, which could best be explained by a similar episode, but a more recent one, half its age or less. These Forster attributed to the Inuit and Na-Dené populations. However, rather than returning to Greenberg's three discrete journeys, Forster saw this double episode as part of a global phenomenon, nested within the earth's climatic history. His two episodes lay either side of the considerable expansion of ice from the poles. Like other northerly peoples, the populations of ancient Beringia expanded when the climate allowed it, and contracted when it did not. Furthermore, Forster's climatically driven account cast a new light on the genetic 'arcs in the sand'.

waves of migration or depletion?

The wave-front model of migration and its relation to 'arcs in the sand' would work better if particular genetic types were found to be absent *beyond* the arc. That would indicate that a particular lineage only got so far. What we find instead is that lineages may be rarer or absent *in the lee* of the arc. The C lineage is less frequent in the north, and two lineages seem to be completely absent along the presumed pathway from Asia into America. One of these is lineage B; the other, known as lineage X, is one of the rarer lineages that had begun to turn up. In both cases, some had speculated on quite separate journeys to account for their complete absence from the northern land route. Could lineages B and X have nothing to do with the northern bridge and instead have reached America by sea?

The first ancient DNA data from pre-Columbian America could in fact have supported the idea of arrival in part by sea. These early pre-Columbian amplifications yielded only A, C and D lineages. When Erika Hagelberg's study of South American mummies finally came up with the nine base-pair deletion characteristic of lineage B, a short-lived but romantic notion of ancient Pacific island journeyers began to recede from the account of American settlement. Since then, a significant proportion of the 500 or so pre-Columbian individuals examined have proven to be of lineage B, including an 8,000-year-old hunter stranded 3,000 metres up in the Rockies. The B lineage was clearly an early overland arrival, much as Forster and others argued, which had disappeared from the path of its entry. This left the yet more enigmatic lineage X.

Merriwether had not been alone in finding lineages that fell outside the classic four lineages of Wallace. One recurrent outsider became labelled 'X'. It was characterized by mutations along the control region at positions 16,223 and 16,278. In one native American community, the Ojibwa, as many as one in four were of this rarer lineage. Like lineage B, it was absent from its presumed ancestral path from Siberia across to Alaska and Canada. What is more, this lineage seemed to be absent throughout Asia. The first geneticist to recognize this lineage, Antonio Torroni, had found the lineage, not in Asia, but in various parts of Europe. Could it be that some forgotten European travellers had carried the lineage across the Atlantic Ocean? Not according to Forster's calculations using the molecular clock. These placed lineage X way back in prehistory, well before such a journey was feasible. Once again, the ancient DNA might cast light on this issue. Most ancient DNA studies of pre-Columbian Indians had targeted too small a sample to pick up rare lineages. There were however some important exceptions.

the pre-columbian 'oneota' of illinois

Around 200 years before Columbus arrived, a short-lived village community buried its dead within a mound on a bluff above the Illinois River. They were from a culture archaeologists describe as the 'Oneota', related to today's Sioux. Most communities of the Oneota culture inhabited the Central Plains, further to the west. This particular village was an outlier, and perhaps suffered as a result. Many of the buried bodies had suffered a violent death, and their burial mound consequently conserves perhaps just a generation or two of their entire village. During the 1980s the burial mound was threatened by a highway development, and a rescue excavation was mounted by the Illinois State Museum. The 260 skeletons they uncovered were made available to Mark Stoneking and his student, Anne Stone. The work they did on the skeletons stands as a model study of the genetics of an ancient population.

Virtually all of Stone's ancient Oneota population was found to belong to one of the four Wallace haplotypes. This is also true of the several hundred pre-Columbian burials that have now been typed, spanning the entire New World from Hudson Bay in the north to Tierra del Fuego in the south. What they reveal is a pattern and geography of pre-Columbian mitochondrial haplotypes that closely match those among living Native Americans.

Like Peter Forster, Anne Stone went beyond a study of well-known lineage markers to look in detail at variation in the mitochondrial control region. This is what allowed her to pick up extra lineages among her sample. She realized that two of her Oneota individuals came from an unusual lineage, which she first published as 'lineage N'. On seeing her publication and reading the mutation site involved, Peter Forster realized what she had. These pre-Columbian individuals were of lineage X. The rare lineage possessed by 25 per cent of the northern Ojibwa people, who live today just a few hundred kilometres to the north of where the Norris Farm mound was excavated, could also be found among that same Norris Farm population 700 years earlier.

Norris Farm may provide the most detailed glimpse of pre-Columbian lineage X, but it is not alone, nor is it the earliest example of the lineage. It has also been found among Brazilian clay pot burials which are 4,000 years old, and among the bodies recovered from the Windover Bog in Florida, which are 8,000 years old. These findings all support the genetic argument of Peter Forster and others, that the lineage is ancient. It entered America before a sea journey was plausible, and must have come across the northern land bridge. Lineages B and X illustrate how the great climatic fluctuations of the Quaternary Epoch not only determined the schedule of population movement, through exposing land bridges, but also wiped parts of the slate clean by eradicating lineages from the harsher environmental regions.

One cannot help comparing Greenberg's three journeys in prehistory with Columbus's three ships sailing several thousands of years later. In each account, there is a hazardous journey to be made between the Old World and the New, followed by rapid infilling of the continent by the journeyers' descendants. Perhaps the concept of a perilous journey has been a distraction, one of several 'great journeys' that have rather over-simplified the human past. The data discussed above are not best explained by three small and discrete migrations, grabbing windows of opportunity to make a dry crossing of the Bering Straits. They fit better with a rather larger number of people coming across, from diverse ancestral homes. We have also had to rethink the timing and geography of any genetic bottleneck. From a modern-day, interglacial perspective, it is the crossing of the frozen north that constitutes the obvious bottleneck, the obvious challenge to expansion. However, from the standpoint of a late Pleistocene fisher-forager, whose family and dogs had lived for countless generations upon the windswept plains and grassy ranges of the vast land that now lies beneath the sea, there was no real sense of crossing from anywhere to anywhere. Instead that ancient population was fluctuating and responding to long-term climatic change. Some periods were ones of expansion, when lineages would be found further to the south. Other periods were ones of depletion, in which lineages would diminish and fade from some regions completely. The gradual population of what would later be transformed into a separate continent was imperceptible. In a sense, the molecular evidence has caused the concept of a great journey more or less to disappear.

ocean crossings

If the journey across the frozen north has been rewritten, what about the ancient seaward journeys that captured the imagination of Grafton Elliott Smith and many after him? The charters of arrows across the ancient map were absorbed by what to do about vast stretches of ocean. This was once an issue for anyone seeking connections between the Old World and the New, although we now know what a lowered sea level can do to the Bering Straits. It remains an issue for Australasia, which, even at the lowest sea levels, is still a seventy-kilometre sea journey from Asia. It became pivotal when envisaging the kind of intercontinental contact and world-wide island-hopping argued by followers of Smith's global hypothesis.

It was with that in mind that Thor Heyerdahl set out in 1955 on one of several ocean voyages, this time headed for Easter Island in a remote part of the South Pacific Ocean. In the tradition of Smith, he was intrigued by the idea of a global diffusion of civilization with its roots in ancient Egypt. He wanted to know whether civilization could have been brought from Egypt by seafarers to Central America, whether it could have travelled from Central America southwards to Peru and, from there, across the oceans to the remote Pacific islands. He had established a few years earlier, in his famous Kon-Tiki voyage, that the second sea-leg was technically possible with a simple balsa-wood raft. Fifteen years later he would do the same thing for the first leg in the second Ra voyage. Now he was travelling to Easter Island in the relative comfort of a Norwegian trawler with a team of archaeologists hoping to recover evidence of the early voyagers themselves.

His team set to work on several parts of the island, surveying and excavating both the ancient remains of houses and the areas around the massive stone statues for which the island is famous. These were the monuments that had brought thoughts of prehistoric voyagers to mind, 'Children of the Sun' as Smith's colleague, William J. Perry, had dubbed them. These stone statues and the stepped altars on which they stood each proved to be part of a monument of many phases. They began as simple altars in the first millennium AD, and ended several centuries later as well-worn ancient sites, with human burials incorporated into the rubble. These buried skeletons, and others that were excavated from separate graves nearby, provided a glimpse of the direct descendants of the first voyagers, before later Polynesians and Europeans had arrived. They were not the relatives of today's Easter Islanders. Since Europeans discovered the islands in the eighteenth century, there has been a fair degree of population movement and replacement, and to the living islanders the great statues are as foreign and ancient as they are to us. By the time their families had reached the island, those who venerated the statues had died out. But the bones from Heyerdahl's excavations belonged to people for whom the statues were part of their living world. These were the people for whom Grafton Elliott Smith's diffusionism had suggested a South American ancestry, an ancestry that would save the notion of a single heartland of civilization. The bones, however, could neither speak nor offer any clue as to their ancestry – that is until their molecular tachometers were recovered and decoded.

Thirty-four years after the excavation, Erika Hagelberg was on her way to Chile's Museum of Natural History where those human remains had been stored. Only a few months earlier *Nature* had published her groundbreaking paper, which presented evidence for the survival of DNA in ordinary archaeological bone. She was keen to take on an archaeological problem, and here was one. Did the ancestors of these ancient people travel in simple wooden rafts across the Pacific from America as Heyerdahl had surmised? If they had, then any remaining fragments of their mitochondrial DNA should trace their path. A decade had passed since the human mitochondrial genome had been mapped and, by now, a fair amount was known about regional variations in the setting of that sensitive evolutionary stopwatch, the mitochondrial control region.

Hagelberg succeeded in amplifying a stretch of the mitochondrial control region from some of these skeletons, and three particular base positions caught her eye. Among living people on the Asian side of the Pacific, one particular haplotype was recognizable by deviations from the Anderson sequence at positions 16,217, 16,247 and 16,261. This was not a haplotype found in the New World, and so provided a useful marker of ancestry. Sequences amplified from the ancient Easter Island skeletons displayed replacements at all three positions, sharing this pattern with populations found on only one mainland flank of the Pacific. Their ancestors had sailed from the direction of Asia, not America.

from balsa-wood boats to 'fast trains'

Heyerdahl is a romantic at the tail end of the Smith and Perry tradition. He also knows a great deal about seafaring and the Pacific Ocean currents. With this knowledge, he was undoubtedly able to raise the debate about ocean crossings above what had been a poorly informed perusal of vast stretches of blue on a flat map. The path of the 'Children of the Sun' was, however, already fading from view. Around the time that Heyerdahl was making his sea voyages from America, others were piecing together the archaeology of the islands further to the west and, like Heyerdahl, making use of the newly developed method of radiocarbon dating. From New Zealand, Jack Golson was beginning to make sense of a particular style of stamped pottery recognized on a New Caledonian beach called 'Lapita', and found on other islands as far afield as Fiji and Tonga. Charcoal on the Lapita beach had provided dates for the pottery, placing it in the early first millennium BC. The geographical spread of the Lapita pottery fitted neatly at the heart of a larger region in which all the modern islanders speak related languages and have similar words for a range of plants and animals.

Here were the makings of a Lapita culture and an alternative set of ancient journeys into the Pacific, originating from the Old World in the west, rather than the New World in the east. As excavations of the settlements with Lapita-style stamped pottery got under way, fragments of their fishing, farming and gardening lives came to light. The islanders kept pigs and fowl, and repeatedly used their considerable seafaring skills for fishing and exchange with other distant islands, as well as to settle new islands. The growing number of carbon dates for these sites fell within a 1,000-year time band. On the islands just north-west of New Guinea, dates as early as 1500 BC were recorded. The carbon dates came out as progressively more recent as one moved to the south and east, indicating that by 500 BC the settlements had spread over 5,000 km into the heart of the Pacific Ocean. The pace of this expansion has attracted the nickname the 'fast train to Polynesia'.

The significance of this prehistoric fast train from the west was enhanced when it was placed in the context of contemporary evidence of language and human genetics. The languages spoken in the Pacific islands cluster within the 'Austronesian' language family, a group whose greatest diversity is found in Taiwan. If a tree is built from variations in these languages and used in much the same way as a genetic tree, then it traces a clear arc. From its root in Taiwan, this arc branches southwards and eastwards towards the increasingly remote islands of the South Pacific. The genetic evidence is more complex, with a clear influence from island Melanesia as well as Taiwan and mainland Asia. Nevertheless, from blood proteins and DNA sequences alike, a restricted range of twigs on the Asian branch of the mitochondrial family tree can be traced as we move towards the remote islands of the Pacific. Modern languages and modern genetics described a sweeping arc across the Pacific, with, at its centre, a smaller arc derived from ancient evidence, the dated sites characterized by stamped Lapita pots and domestic animal bones. Peter Bellwood brought these together as evidence of an ancient suite of voyages by pioneer farming settlers. He suggested that they set out from Taiwan as much as 4,000 years ago and reached the most remote islands of all around 1,500 years ago. On the way, they laid the foundation of the Lapita culture in the central part of the arc.

Hagelberg's ancient Easter Islander DNA gave a much better fit with this set of journeys. In a later comprehensive survey of the mitochondrial control region among modern islanders, Bryan Sykes found all three of her key base substitutions right across the arc, and two of the three substitutions could be traced back to Taiwan. Here was a pattern she could explore by extending her study of ancient bones on Easter Island to some even older bones elsewhere in Polynesia. With a more comprehensive range of ancient islanders and their DNA, she could really get to grips with this 'fast train' from Asia, or so she assumed.

Her work on the other Polynesian islands involved a characteristic mitochondrial feature that was playing a large part in piecing together the earlier journeys of colonization of the New World explored above. This was the deletion of nine base pairs on the far side of the mitochondrial genome from the control region. It was the same deletion that Douglas Wallace had used to identify lineage B among the Native Americans. This deletion of nine base-pairs was also encountered in the islanders of the South Pacific, where it was extremely common. On the continental mainlands on either side this deletion was held by a significant minority of individuals, but here it was carried by a sizeable proportion of Pacific islanders. Of the modern Polynesians that Sykes surveyed, 94% carried the deletion. Hagelberg realized that she could use this deletion, together with the control region variations she had found at Easter Island, to explore the alternative journey and trace the path of pioneer farmers along the arc.

Ancient skeletal remains were collected from around a dozen Pacific islands, and DNA successfully amplified from twenty-one of them. Eleven of these were just a few centuries old, going back 700 years at most. Nine of those eleven bodies carried the nine base-pair deletion and, where the amplification was successful, the control region generally displayed the characteristic replacement of three bases. These ancient islanders were part of the same lineage as both modern islanders and the ancient Easter Islanders. However, none of these corresponded to the actual users of Lapita pottery. To explore the earliest part of Peter Bellwood's story much older skeletons were needed. Lapita burials are not common, but Roger Green from New Zealand had found and excavated a small cemetery with eight adults on Watom Island. Hagelberg gathered a further ten skeletons, between 1,700 and 2,700 years old, from the cultural context in which Lapita pottery was used. They may have used the pottery, but their mitochondrial sequences were not what would be expected of the 'fast train to Polynesia'.

None of the ten earlier skeletons carried the nine base-pair deletion. What is more, the three from which the control region was amplifiable displayed virtually no replacements on the characteristic sites. The people who actually used the Lapita pottery appeared on the face of it to be different from modern Polynesians. In spite of the geographical match between the ancient Lapita evidence and the modern variation, and some elegant archaeological arguments in favour of continuity between the Lapita culture and more recent material culture traditions, the ancient DNA evidence suggested a break in genetic continuity.

Everyone, including Hagelberg, was cautious about making too much of a handful of skeletons. Six per cent of modern islanders, like the Lapita burials, lack the nine base-pair deletion. Perhaps a rather larger sample of ancient sequences will modify the picture and bring the ancient islanders back into the genetic fold. However, even small samples are informative. Imagine tossing a coin a small number of times. If it falls head upwards two or three times in a row, that can be put down to chance. If it falls head up ten times in a row, then it is not unreasonable to suspect that the coin might be weighted. By analogy, the most straightforward explanation for the absence of the very common deletion from all ten skeletons is that the material culture traditions over-simplify the demographic history of the islands. There was a series of journeys, not all from the same starting point, that have accounted for the islands' various populations through time. There may be some continuity between the material traditions identified on the Lapita beach and those of later times, but different human lineages with different ancestries became the agents of that continuity.

The Pacific islands provide an interesting example of how the retracing of ancient journeys has proceeded. Starting from evidence from the present day, be it of genetic traits, language or surviving monuments, projections back in time towards their origins lead us towards some fairly simple stories, stories that can easily be portrayed by sweeping arrows across the map. That is to some extent inevitable. Given a choice of conjectural paths that might conceivably lead to the present, it is good scientific practice to choose the simplest explanation. As scientific dates first, and then our molecular tachometer have come on the scene, we might hope they would select which simple explanation is the correct one. What they have instead tended to show is that real life is more complicated than that, and simplicity is not a particularly relevant criterion in tracking real human pasts. Hagelberg's work on ancient DNA provided a neat fit for neither the eastward nor the westward journey in their simplest forms. There were other things going on, and perhaps rather more journeys involved, something borne out by Bryan Sykes's extensive survey of the modern DNA.

Of more than 1,000 modern Polynesian islanders surveyed, the great majority of them fitted in with the idea of a fairly tight genetic bottleneck letting through a narrow suite of mitochondrial haplotypes. However, a couple of individuals stood out. One individual from Tahiti and another from the Cook Islands had sequences that were unusual for Polynesia. The best matches were to be found in Chile and among the Mapuche Indians from Patagonia. Perhaps there was an ancient journey following the Kon-Tiki path, even if it was not the pioneer journey. Perhaps more women were braving the ocean waves than at first thought. They would need to be women to carry the mitochondrial lineages. When we move to the men, the journeys really proliferate.

the gauguin effect

If we were really desperate to resurrect the Children of the Sun, we might muse on the similarity between the unusual mitochondrial haplotype from Polynesia and a closely related sequence from the Basque country in western Europe. Sykes makes only passing mention of this rather peripheral possibility, but when he and his colleague, Matt Hurles, shifted their attention from the maternally inherited mitochondrion to the paternally inherited Y-chromosome, Europe loomed large. What they found was that the Y-chromosomes from no less than 30 per cent of modern Polynesian men descend, not from any inhabitants of the adjacent continental landmass, but from distant Europeans. The sailors, whalers, artists and missionaries, at whom Sykes and Hurles waved a stern finger, traced a varied pattern of discrete paths into the Pacific, and yet the genetic influence has been substantial. They almost certainly occupy quite a late chapter in the story, but illustrate a central element of the whole tale.

civilization comes home

Ancestral journeys and cultural identity often go hand in hand. It suits us to know where we have 'come from', and the stories we tell about that journey of settlement help to forge a sense of collective identity. Essential elements of those stories are the places we definitely did not come from, inhabited by people from whom we see fit to differentiate ourselves. The notion of an ancestral journey is more than a matter of dry, scientific record. It has also played a major role in forging many a cultural identity, not least in the continent that has delved deepest into its own prehistory to trace the journeys at the foundation of western civilization.

For centuries, European scholars have been intrigued by the civilizations of antiquity. They have told a number of different stories about how those ancient civilizations came about and how they led to the modern civilizations of which they themselves were part. The routes to building those stories were widespread. Researchers scrutinized ethnic traditions and notions of race, and looked at the variations between different European languages and certain Asian languages whose vocabularies seemed to display similarities. They examined archaeological remains and posited ancient territorial expansions that resonated with their own experience of European territorial expansion around the world.

In the nineteenth century, those stories had at their end point a white, Christian, civilized Europe, increasingly compartmentalized into nation states. Their starting points were the ancient archaeological sites that seemed to reveal so many of the precursors to European civilization – towns, trade, monuments and the written word – features that drew attention to the east Mediterranean and beyond, to Mesopotamia. Around this civilized core were different cultures, described by words such as 'barbarian', and generally associated with darker skins, simpler settlements and the absence of a written text. In telling stories of the origin of white civilized Europe, it was as important to keep these peoples and their cultural traditions outside those stories as it was to keep other peoples and traditions inside. As the prehistories of north and east Africa became better known, so would this exclusion require an increasingly vigorous editing job.

In his book *Black Athena*, Martin Bernal examined the editorial task facing nineteenth-century scholars in diminishing the place of Africa in the story of Europe. The greatest challenge in this was the Nile Valley, an obvious corridor of contact between the two continents. Those scholars knew of the evidence, both textual and archaeological, for influences upon Greece from Egypt and the continent to the south. This tended, however, to be put to one side in favour of evidence of influence from the east that was in line with the more palatable Aryan stories of origin. The growing study of Indo-European languages provided a new basis for charting these Aryan journeys, and marking out boundaries in tune with contemporary notions of European identity.

The various data-sets, archaeological and linguistic, that were brought together in the context of these nineteenth-century ideas remain with us, and form the starting point of our inquiries into

Europe's past. It occurred to a number of the new molecular archaeologists that the more contentious ideas about Europe and its neighbours could be tackled with modern and ancient DNA, in particular the relationship between ancient Europe and Africa. It should be said straight away that such a relation has two dimensions, only one of which is accessible through DNA. The interchange of ideas and cultural concepts must continue to be explored through text and artefact. It is only when people too are moving or intermarrying that something may show up in the molecular tachometer of DNA.

journeys from sub-saharan africa

When it comes to tracing this particular journey, molecular archaeologists have two points in their favour and one major point against. The first point in their favour is the rich diversity of African genetic sequences. Each of the human family trees assembled from molecular information has revealed a considerable breadth in the continent where humanity evolved, with only a reduced range of genetic diversity leaving Africa to populate the rest of the world. This provides a range of haplotypes to serve as specific sub Saharan markers. The second point in favour is the long history of mummification, both natural and artificial, along the Nile Valley, the principal corridor between Africa and the Mediterranean. Indeed, it was an Egyptian mummy that provided Svante Pääbo with his first identifiable sequence of ancient human DNA. Set against these two positives, the major negative is that much of the Nile corridor has been, and remains, extremely hot – not the ideal conditions for the preservation of ancient biomolecules. None the less, it was worth a try.

Midway between Egypt and sub-Saharan Africa, the Nile passes through the ancient kingdom of Nubia. The Nubians were known to be in contact with the societies to the north and the south, and were geographically on the crossroads between distinctive human gene pools. Several population movements up and down the Nile are recorded in history and the DNA of the people living in the Sudan today will reflect that. In Barcelona, Carles Lalueza Fox wanted to reach back beyond those more recent migrations to see what movement there had been by the time of the classical civilizations of the Mediterranean. A Spanish archaeological team in Sudan had been excavating near the modern settlement of Abri, just over 100 km south-west of Lake Nuba. They had unearthed a cemetery of naturally mummified bodies, radiocarbon dated to the later first millennium BC, providing Lalueza Fox with some suitable specimens for ancient DNA research.

For a genetic marker, he chose a point mutation that appeared to be very ancient indeed. It had occurred prior to some of the earliest branching in the human mitochondrial tree, in a part of the tree that remained within Africa. The single base replacement at position 3,594 on the Anderson sequence is found throughout sub-Saharan Africa, but is now rare elsewhere in the world. It is carried by over 95 per cent of Pygmies and San and !Kung bushmen, and by around 70 per cent of sub-Saharan Africans as a whole. The mutation provided Lalueza Fox with an ideal tracer of movement northwards of sub-Saharan genes. His ancient Nubians lay midway between sub-Saharan populations and those in Egypt and the north. He cleaned and ground teeth and bones from twenty-nine ancient individuals, and amplified a 109 base-pair sequence spanning the mutation site. Fifteen of the individuals came up with a successful amplification, revealing that four of the fifteen carried the marker mutation that led back to sub-Saharan Africa. That 25 per cent is midway between the figures for populations to the south and to the north. Looking at these results statistically, Lalueza Fox concluded that somewhere between one-fifth and a half of the population's genes derived from sub-Saharan Africa, and the remainder from the north.

To look at the north-east African corridor more closely, he went on to place the ancient Nubian data in the context not only of modern data from Ethiopia and Israel, but also of the continental data from Africa and Europe. The characteristic mutation was carried by almost three-quarters of those from sub-Saharan Africa, just less than half the Ethiopians examined, around a quarter of the ancient Nubians, one in eight Palestinian Arabs, and a mere trace of mainland Europeans. In antiquity, it was not just political influence and trading goods that were moving along the Nile corridor. The data fell along the corridor in a neat 'cline' or linear sequence of variation. The genetic cline shows that people were moving back and forth as well, and in significant numbers. Since then, Matthias Krings, who in Pääbo's lab had led the Neanderthal analyses, has also published data on modern variation in this particular sub-Saharan marker. Mainland Europe lies at the end of this cline, and the frequencies for the characteristic mutation fall virtually to zero, but not quite.

Throughout history, the island of Sicily has been one of the major cultural crossroads of the Mediterranean. Sure enough, a survey of ninety unrelated islanders produced a frequency of the characteristic sub-Saharan mutation at just over 4 per cent. This is about a third of the frequency at which the mutation is found among the only Arab data then available, from Palestinian Arabs studied in Israel. There are many episodes in the past when this gene flow could have linked the central Mediterranean ultimately with sub-Saharan Africa, and the mediaeval period of Islamic Sicily is an obvious contender. Indeed, the untested assumption when African haplotypes turn up in modern European populations is that they relate to a recent episode of immigration. However, with so little ancient human DNA study, placing the flow within a time frame remains a matter for conjecture. What the ancient Nubian mutations do indicate is that a substantial flow of genes was occurring between the south and the north of the Sahara at a time when the classical civilizations of the Mediterranean were flourishing.

ex oriente lux

For the ancestry of Europe and its civilizations, scholars have repeatedly looked to the east rather than the south. Their points of focus have been Middle East and Near East as the source of all things European, a principle summed up in a phrase that captured Childe's imagination – *Ex Oriente lux*, or 'Out of the Orient, light'. In the arguments of different scholars, significant movements of peoples from the east have clustered in one or more of three phases. A very ancient phase is associated with Palaeolithic hunter-gatherers, a subsequent phase with pioneer farmers, and a still later phase with monumental sites and the knowledge of metalworking.

Colin Renfrew had used carbon dates to query the extent of movement in the third of these phases. By the time the mitochondrial control region had been recognized as a rather interesting evolutionary tachometer, Cavalli-Sforza's emphasis on the second phase, based on the evidence of expressed genes, was gaining adherents. Little interest had been shown during the 1970s and early 1980s in his demic diffusion argument for a burgeoning population of pioneer farmers expanding from the Fertile Crescent. It had not really struck a chord with archaeological thinking at the time. By 1986, however, his genetic database had reached impressive proportions. Furthermore, it was backed up by a comprehensive archaeological and linguistic survey. In 1987, Renfrew published a detailed argument to account for the spread of Indo-European languages. He attributed that spread, not to the later prehistoric horse-riding Kurgan cultures favoured by Maria Gimbutas, but instead to a wave of pioneer farmers, drawing directly on Ammerman's and Cavalli-Sforza's demic diffusion model. That 'wave of advance', spreading from the south-east across to the north-west, now seemed to enjoy the support of three quite distinct data-sets, and notable support from a range of disciplines. In 1994, Cavalli-Sforza's global synthesis of human genetic variation was published. It was a truly exciting work, in which thousands of years of historical geography were revealed in the minute variations in the blood of the living. The great journeys of the human past could be followed over every continent. But within a couple of years another line of genetic evidence was telling a different story.

While long lists of blood group data were being brought together with other genetic information by Cavalli-Sforza and his team, a much smaller fragment of the human genome was under scrutiny by Bryan Sykes's team in Oxford. They were engaged in the now familiar task of charting the individual variations in the control region of the human mitochondrial genome. It was an approach that had already been taken in Asia, America and Polynesia. Now it was being applied within Europe. Samples were acquired from just over 300 modern Europeans from as far afield as Portugal and Finland, together with samples from the Middle East and Turkey. One of the most variable stretches of the control region was sequenced in each case, and the sequences compared. Although the team was working with one of the fastest evolving sequences of the human genome, the 10,000 years since agriculture began remains an extremely short period of time. We would not expect a great deal of diversity to have accrued since the founder population spread across the continent – perhaps one or two base replacements in some individuals. Sykes and his colleague, Martin Richards, found that the variation was rather greater than that. Almost a third of the individuals studied deviated from the Anderson consensus sequence, some by as many as six base replacements along the short stretch of targeted DNA. This European group was looking rather more ancient than a population passing through a very recent bottleneck. Sufficient time had passed to allow a significant amount of evolutionary change. As we have seen with the colonization of the New World, however, this need not preclude a simple model of population expansion. Just as the ancient Beringians could bring a fair degree of diversity with them on their passage into America, so could the ancient farming populations from the Fertile Crescent.

This was where Sykes's Middle Eastern data came into play. If they had brought that diversity with them, it might still be discernible in today's populations from the Fertile Crescent. The forty-two Middle Eastern individuals in Sykes's sample displayed something different. Rather than encompassing the European variation, they formed instead a rather distinctive group, out on a limb from the broader European pattern, and sharing their sequence pattern with only a small minority of modern Europeans.

Sykes and Richards realized that their targeted analysis of one well-studied stretch of DNA was telling a quite different story from what had emerged from Cavalli-Sforza's masterly survey. They argued that, while a population spread of early farmers from south-west Asia had taken place, it only contributed a small fraction to the European gene pool. There was no wave of population advance sweeping across Europe, just a trickle. The sequence diversity had to be explained in terms of a more ancient population. Their hypothesis cut across the core journey within Cavalli-Sforza's grand synthesis. The principle ancestors of modern Europeans were the hunter-gatherers who were already there, thousands of years before the spread of farming. Agriculture travelled as an idea more than as a community, and the presumed population replacement had not taken place.

A few months after the publication of the Oxford findings, Cavalli-

Sforza delivered the McDonald lecture at Cambridge. He projected one synthetic map after the other, taking a captivated audience through several major episodes of the human past. We waited to hear his assessment of the Oxford paper, but when it came his characteristic charm and ease were put to one side. His feelings were unambiguously expressed first towards the Oxford group, and then towards the offending journal for letting their flawed arguments through to publication.

While a polite Cambridge audience wrestled with a certain amount of *schadenfreude* at so eminent a disposal of an Oxford project, we were left wondering how they had managed so to rattle the senior man. It would have been less surprising to see him dispose of them with a learned smile, rather in the manner of dispatching a careless question or overly arrogant student. The challenge clearly ran deeper than that, for all the implications that the Oxford approach was carelessly flawed.

The analysis put forward by Sykes and Richards was not only different from Cavalli-Sforza's approach in its conclusion, but also in its conception. Cavalli-Sforza had been concerned with the larger genetic picture created by extracting patterns of similarity in the expression of between 50 and 100 genes and looking for parallel patterns within linguistics and the archaeological record. The approach taken by Sykes's team was much more narrowly targeted. The DNA sequence they examined was far shorter than even one of the many genes Cavalli-Sforza utilized. He had sought out global trends within a vast genetic data-set, while Sykes had sharpened the focus on a specific evolutionary process involving only part of a single DNA molecule. In terms of an analogy with exploring the evolution of flight, one scholar was looking across the skies at a multitude of airborne forms, while the other was dissecting the wing of an individual bird. The different styles of their conclusions were in part a consequence of that.

At first, academic opinion tended to side with Cavalli-Sforza, and various alternative reasons were put forward to explain away the Oxford group's results. It was pointed out that the mitochondrial sequence is only a very tiny element in the human genome and should not provide a basis for argument on its own. Others pointed to the potential skewing effect of residual hunter-gatherers, reflected in the language and genetics of today's Basque and Saami peoples. A further weak point was the mitochondrial clock, the powers of precision and accuracy of which are habitually over-used. It could always be argued that such a timepiece might vary in a way we simply do not understand, and in certain cases be several magnitudes out. Bryan Sykes, who had stuck with the work and enhanced his results, could see that the way to move forward was to look in well-dated, well-positioned contexts for ancient DNA.

What he needed to do was to find examples of the potential ancestors in each story – ancient European hunter-gatherers who would serve as indigenous ancestors, and early Neolithic farmers as part of the pioneer population advance. At London's Natural History Museum, Chris Stringer found Sykes some indigenous hunter-gatherers from a limestone cave in Cheddar in the south-west of England. These caves retained many remnants of the hunters and gatherers who had sought shelter here several thousand years before the first cereals and domestic livestock reached Britain's shores. Alongside the abandoned hearths and the disused flint points from spears and arrows, the bodies of some of their users remained. Sykes examined two of the human skeletons from the caves, 13,000 and 9,000 years old respectively, and was much reassured by what he found. They were more than plausible ancestors of those still living in the region today. He actually found individuals, living locally, who carried precisely the same control region haplotype as the unearthed hunter-gatherer who had lain at rest for so long in a nearby cave.

The Cheddar cave individuals strengthened the argument for a local, pre-agricultural ancestry. What Sykes needed now was a group of pioneer farmers who corresponded with the population trickle that did get through, bringing with them the ideas and traditions of farming. One of the best candidates in the archaeological record for such a population movement is the so-called 'LBK' culture. LBK stands for *Linear Bandkeramic*, the linear banded pottery that characterizes a series of early Neolithic sites scattered across the loess soils of Europe north of the Alps. The sites share not only a particular form of pottery, but also a typical cluster of timber long-houses, a particular range of domesticated plants and animals, and even a characteristic set of weeds accompanying the crops. The acid loess soils are hardly the ideal place for the kind of bone preservation conducive to ancient DNA studies. By 1996 the Oxford group were looking for some of the first generation of farmers of temperate Europe to bring domesticated crops and livestock north of the Alps, in order to establish what haplotypes they had brought with them. They visited three sites, two in Germany and one in France, while these were being excavated, so that they could extract bones carefully and take them back to the lab. Of nineteen individuals examined, only three provided clear evidence of native, uncontaminated DNA, but these three were indeed interesting. In the stretch of the control region studied, one of the three followed the consensus sequence, while the other two displayed mutations at positions 16,069 and 16,126. This takes our LBK pioneer farmers out on to a particular limb of control region variation, precisely the same limb as that occupied by modern populations from the Middle East. The contrasts between the hunter-gatherer from Cheddar Cave and the LBK farmers seemed to resonate with Sykes's and Richards's account. There was indeed a gene flow from the south-east, linked with the spread of farming, and the Middle Eastern genes flowed most discernibly with particular populations, such as the LBK farmers. Among those who eventually turned to agriculture, the great majority of European farmers trace the larger part of their ancestry back to the likes of the hunter-gatherers from Cheddar Cave.

The spread of the new farming resources was subtle and in many cases quite gradual. Most of those who learnt to till the ground and sow seeds were direct descendants of communities already living in the region, whose intimate knowledge of Europe's varied environments their descendants carried through alongside their new farming skills. The last major swathe of human migrations across Europe had probably already happened.

which were the great journeys?

Cavalli-Sforza and Sykes have now reached an accommodation of each other's models. Each has acknowledged the past complexity that can accommodate both data-sets. An easy way of presenting the debate that went on between them is as a contest between two conflicting ideas. One highlighted a swathe of ancestral farming journeys, transforming demography, culture and language on a massive scale. The other emphasized people who doggedly stayed put, picking up new ideas and lifestyles, but barely moving a stone's throw from the ancestral cave. That would be a misrepresentation. The work of Sykes and several others who have combined modern and ancient DNA studies has identified a variety of journeys. However, they are journeys that vary considerably in scale. Some involve the wholesale colonization of an unpopulated continent or island group. Others involve intermediate trickles across Europe or along the Nile corridor. Then there are the uncharted exploits of a few South Sea whalers. These journeys also vary considerably in how they match up, or indeed how they often fail to match up, with the parallel movement of ideas, language, material things and ways of life. Global patterns there may be, in material culture and genetics alike. However, when we unpack those patterns and attempt to trace particular lineages in detail, we do not see great convoys. We do not encounter pioneer settlers, on horseback and in sea boats, carrying with them blueprints for establishing whole new ways of living, while the previous occupants fade into oblivion. We see instead a complex assortment of meetings, melting pots and mixtures, people changing the way they do things, and embarking on smaller journeys that map directly on to neither language nor culture.

How much is this busy picture in conflict with the smooth clines of Cavalli-Sforza's grand synthesis? His principal components analysis did not paint a picture of unbounded complexity; there clearly have been significant and coherent trends in global gene flow, and those require an explanation. None the less, a trend is not a migration, an expansion or a journey. It is instead a pattern that emerges from a series of processes that deviate from randomness in a convergent manner. For example, the north-west to south-east cline in European genes indicates an axial trend in gene movement, but there is no necessity to conclude that this is the consequence of a single episode. It could result from very many movements in both directions, brought together by a shared axial tendency. While the mitochondrial control region may show the great variety of movements in the human past, Cavalli-Sforza's gene maps indicate how in the longer term those movements converge upon particular trends. These trends can furthermore be related to parallel trends in language and material culture. Indeed, if we go back to look closely at the original use of the demic diffusion model by Cavalli-Sforza in the 1970s, we find there is no need for them to imply coherent migrations in any form. The outward movement of genes is an emergent consequence of a constellation of random trajectories, not of any sustained journey along a particular path.

8

beyond DNA

the other molecules

The molecule at the heart of life has cast a fresh light on past human lives. Its beam has ranged from pre-contact America, to the Nile Valley at the time of the classical civilizations, and further east to ancient communities in Siberia, China and Japan. It has shed light upon prehistoric farmers in Europe, central Asia and the Andes, and on the cereals and livestock on which they depended. That same beam has reached back in time, to Palaeolithic hunters and to a range of animals that have since been hunted to extinction. Further back still, it has illuminated the genetics of our extinct relatives, the Neanderthals. The glimpses we have caught of this genetic information allow us to look afresh at the movements of prehistoric peoples and their ancestries, and place their stories within a clear archaeological context and time scale. Their stories in the process become reworked and retold, and some of the complexity and detail of human lives is returned to them. But there is still much that is left in the shadows.

Any set of data from the past sets its own limits on what we can see from past human lives. In the first archaeological books I ever encountered, it seemed that little of those lives appeared, beyond a pair of detached hands fashioning the pot or stone tool that at the time was the focus of attention. Bio-archaeology added greatly to that vista and our focus shifted from the hands to the people, the clothes they wore, the food they prepared, the villages they built and the dwellings they furnished. Once we realized that the surviving fragments of the past extended to DNA, it was as if we archaeologists might enter those dwellings, camera and microscope in hand. In that heady atmosphere, we could easily forget the preconceptions we carried with us into those dwellings, preconceptions arising from what we already thought the archaeological record was like. The ancient DNA spotlight was first trained upon the most familiar bio-archaeological tissue, human remains, and moved from there to animal bones and then to the remains of the best known food plants. So studies of ancient DNA have revealed to us something about farming communities, their principal domesticates, and their journeys and expansions across the world. Other lives that were interwoven with quite different food species, and other things that were going on in the prehistoric world, have remained in the dark. But there is more than DNA lurking amidst the dirt of an archaeological excavation, other molecules to be found and decoded. Several of them are both more abundant and more persistent than the molecule at the heart of life. If they too can reveal information in the kind of detail possible with DNA, then they can take our observations elsewhere, to other facets of the prehistoric world.

This possibility was very much in the mind of Geoff Eglinton when we set up the Ancient Biomolecules Initiative in the early 1990s. At that stage ancient DNA research was enjoying prima donna status. In a molecular drama of many parts, it was the billboard star. True to the spirit of stardom, it could be exciting and disappointing in equal measure, and it could both rise to the headlines and conspicuously flop. Geoff Eglinton knew enough about science to see we could not just work with the star performers. We needed the support cast and the production team. To see for certain what molecular information remained among the dirt and the smells of archaeological deposits, we needed to place ancient DNA in the context of a variety of other ancient molecules that were also known to be around. A few obvious candidates deserved attention.

blood, building blocks and enzymes

The great value of DNA as a source of information lies in its long-chain structure, its encoded sequence of molecular links; but it is not the only long-chain molecule in the living system. The many proteins circulating in the blood, together with those that build our soft and hard tissues and those that form the enzymes that steer the body's chemical processes, also possess a long-chain form. They are assembled from a sequence of amino acid units whose order is determined by the DNA blueprint. Soon after the structure of the double helix was discovered, molecular biologists began to tackle the problem of how that blueprint worked. Researchers established that the DNA blueprint was read in clusters of three bases. Each triplet of bases, or 'codons' as they came to be known, would line up a specific amino acid to add to the protein sequence. This was achieved with the help of the cell's 'molecular machine tools', the contorted molecules of RNA, DNA's close cousin. These RNA machine tools would allow the DNA blueprint to be read, codon by codon, and a protein to be built incorporating a corresponding sequence of amino acids. So, for example, two guanine bases followed by one adenine base forms the codon that will enable a glycine amino acid unit to be added to the protein chain. As there are four potential bases at each of the three positions of each triplet of bases or 'codon', there are sixty-four different messages that can be conveyed to the construction process. This is more than enough to cover the twenty or so amino acids from which proteins are assembled, as well as to code the 'start' and 'stop' messages needed to define the boundaries of any particular protein-building gene. In other words, the protein sequence preserves a great deal of the information in the DNA sequence, easily translatable from one to the other, but with the potential to reach much further back in time.

The secret of the ancient protein's survival lies in some close attachments it makes with the mineral component of living organisms. For example, the collagen in bone can survive to be a very ancient protein indeed. At Newcastle, Matthew Collins has spent several years working out how ancient bones are conserved. He examined a 120,000-year-old bone with 80 per cent of its original collagen in place. He estimated that in suitably cold conditions some collagen should survive for a million years or more. Before we get too excited about its genetic possibilities, we have to recognize that it is, unfortunately, a rather monotonous molecule. Other proteins could be much more informative.

bloodlines

We would not immediately expect blood proteins to be great molecular survivors; they are lodged in a fluid which is by nature chemically and biologically active. Freshly spilt blood changes colour before our eyes, and offers little resistance to pathways of transformation and decay. Blood nevertheless reaches throughout the body's tissues, here and there providing certain blood proteins with a more secure haven from breakdown and decay, enclosed within a bone, or even a mummified muscle. What is more, the blood itself offers a highly sensitive system of protein detection that could be mimicked in the lab and thus used to track down proteins that have been reduced to tiny quantities with the passage of time.

This sensitive system within the blood involves 'antibodies', themselves a group of proteins whose biological function is to seek out invasive organisms by recognizing alien biomolecules on their own molecular surfaces. This they achieve by attaching a molecular 'mould' to a particular part of the invasive biomolecule. These couplings can be extremely precise, in that an 'antibody' only attaches itself to an 'antigen' that possesses a highly specific molecular arrangement. As our abilities to build artificial biomolecules grew, so the natural antibody/antigen coupling formed the basis for a series of artificial chemical detectives, designed to seek out molecules chosen by the researcher. These techniques of 'immunology' made it possible to seek out proteins, and indeed other biomolecules, in tiny quantities, and are among the earliest established techniques in biomolecular archaeology.

One series of blood proteins comprises the natural antibody systems themselves. Several of these systems have a relatively straightforward genetic basis. It was soon realized they could be used to assemble the kind of stories about origins and migrations explored in the preceding chapter. Soon after the First World War, Ludwik and Hanna Hirschfeld, a husband-and-wife team, did just that. They had assembled information on ABO blood groups from allied soldiers on the Macedonian front, and combined it with data they had gleaned from German populations. The A and B types formed a cline spreading from north-west Europe (England) by way of the eastern Mediterranean to India, leading them to suggest that 'we should look to India for the cradle of one part of humanity'. Fourteen years later, another husband-and-wife team, William and Lyle Boyd, attempted something similar with ancient Egyptians.

Lyle Boyd first had the idea of detecting blood types in ancient mummies in 1933. She had typed blood groups from saliva and dried blood, so why not dried muscle and ancient bodies? Within four years, she and her husband had published their findings from 122 Egyptian mummies, and from a further 205 mummified bodies from America. A striking feature of their results was the repeated finding of blood group B. Among living Native Americans, blood group B is a predominantly northern phenomenon, but the Boyds were finding it in mummified bodies from the south. They were reluctant to challenge the modern evidence. This after all was well before 'biomolecular archaeology' was a phrase that meant anything to anyone. By the mid-1970s the Virginia pathologist Marvin Allison could be bolder. He analysed the blood types of 140 South American mummified corpses, which spanned the arrival of Columbus and the subsequent conquistadors. There was a clear contrast between earlier and later individuals.

The post-Columbian bodies displayed a range of blood types with a high proportion carrying blood group O. That was consistent with the modern distribution. In contrast, the evidence of pre-Columbian bodies mirrored the Boyds' evidence collected four decades earlier. A, B, and AB blood groups were significantly more abundant in the pre-Columbian bodies. Allison had picked up a small evolutionary episode: some blood groups had fared better than others in the face of diseases that flourished after contact with Europe.

ancient blood: from mummies to mammoths

Around the time that Marvin Allison was re-examining blood groups in mummified humans, Jerold Lowenstein at the University of California was turning his attention to much older mammals. He took the methods a step further by building radioactive markers into his immunological reagents. Once they had sought out their target protein, they would mark it with a beacon of radioactivity. As little as a billionth of a gram of the target protein was sufficient to leave a visible radiation scar on a sheet of photographic film.

By now, several blood proteins had been identified from bones that were a few thousand years old, and it seemed timely to push the analysis back into prehistory. In 1977, a baby mammoth christened 'Dima' gave Lowenstein his opportunity. Dima had been exhumed from the Siberian permafrost at Magadan, and had been dead for 40,000 years. Lowenstein designed antibodies to seek out blood albumins, which are important proteins in the colourless part of the blood, and to compare them with the albumins of a range of living mammals. He found the albumins he was looking for in Dima's thigh. About 1 per cent of the original albumins had survived, and they closely matched those from living elephants.

Like the earlier work on the ABO system, Lowenstein's analyses of blood proteins paralleled and complemented the kind of DNA-based phylogeny that has been discussed in previous chapters of this book. The protein evidence had the limitation of relating to expressed genes only, and provided no guide to the uncoded sequences so valuable in DNA phylogenies. However, it had the advantage of longevity. Dima's ancient DNA was just inside the limits of feasible detection, but its albumin proteins were more abundant than the quantity needed for RIA detection by a factor of a thousand. Ancient DNA analysis could just about reach into the realm of major mammalian extinctions, whereas ancient protein analysis moved comfortably into this area. Lowenstein and his colleagues studied a number of extinctions directly parallel to those explored through ancient DNA. In addition to Dima and other mammoths, they worked on the bones of a mastodon and an extinct sea cow, the soft tissues of a Tasmanian wolf and a quagga, and even the crystallized urine of an extinct pack rat. In these studies, there was a very clear and reassuring match between the two forms of molecular phylogeny, from the DNA and the protein.

dinosaur protein

Blood proteins may have some potential in the future growth in ancient biomolecule research. This may involve moving away from the plentiful blood proteins, such as haemoglobin and the albumins, to the less plentiful but highly informative proteins such as the immunoglobulins. Beyond the blood, yet other proteins surprise us by their remarkable persistence. One of these is, like collagen, tucked away in the biomineral region that can offer the greatest resistance to decay.

Collagen is the principal bone protein, but it is not the only one. A small protein called osteocalcin is the second most abundant protein in bone after collagen. It is a short molecule, made up of forty-nine amino acids, and among those forty-nine are three of an unusual type known as carboxyglutamic acid, or GLA for short. These three GLA units form a very powerful adhesive patch, which holds on tight to the mineral particles of apatite within bone. The GLA amino acid has been detected in very old bones indeed.

Dinosaur DNA may have come and gone from the scientific literature, but the detection of the GLA amino acid in 75-million-year-old bone brings these giants back into the picture. It provides strong evidence that the osteocalcin GLA patch has retained its adhesive strength, and such tightly bonded proteins have a considerable chance of persistence. There may well be other minor proteins lurking in the safe haven of the bone's biomineral recesses, carrying key information about the distant past. Ancient DNA has reached just far enough back to give a glimpse of the most recent episodes in hominid evolution. The longevity of ancient proteins is sufficient to embrace hominid evolution in its entirety, and go beyond to much earlier evolutionary stories. We simply await their identification. Such structural integrity has already been found in another range of biomolecules noted for their extreme longevity.

the energy molecules

If DNA carries the code, and proteins do the work, then other major molecules, the energy molecules, are the ones that fuel the whole process. A consideration of this third group takes us to the most ancient biomolecules of all.

Deep in the earth's crust, our vast reserves of petroleum indicate just how persistent some of these energy molecules can be. Their hydrocarbon backbones persist as petroleum, with many of their chemical side-limbs honed off through millions of years of exposure. In the much more recent deposits of the archaeological record, one of the more striking examples of the persistence of these fatty, oily substances is an unusual soapy substance occasionally found around a buried skeleton. Adipocere, as it is termed, has been recognized since at least the seventeenth century, when a certain Sir Thomas Browne wrote in 1658 of a 'fat concretion, where the nitre of the Earth, and the salts and lixivious liquor of the body, had coagulated large lumps of fat, into the consistence of the hardest castle-soap' around the exposed skeleton of an ancient churchyard burial. As is typically the case with ancient biomolecules, adipocere is not there in its original form. The body's fat has fragmented, at a molecular level, into glycerol, which breaks down quite easily, and fatty acids that, together with alkalis in the soil, form the soapy substance which caught the attention of Browne.

molecular overcoats

Fats are more remote from the genetic blueprints than proteins, and they pass through food in a more intact state. In other words, their variation may not tell us a great deal about the ancient organism we are looking at. Nevertheless, some lipids can tell us a great deal. Take, for example, the shiny waxy surface on the leaf of a plant, its waterproof protection against the elements. Within this waxy coat are lipids that are both highly characteristic of their plant origin and among the most persistent organic molecules of all, albeit in a slightly modified state. These are the cutins, long-chain molecules composed of a sequence of fatty acid units linked together. They are the molecules that endow a leaf cuticle with much of its strength.

At London University, Margaret Collinson and Andrew Scott have been looking at these waxy cuticles on leaves that are up to 300 million years old. Even in these ancient specimens, it is possible to lift off the leaf's waterproof overcoat, a smooth thin sheet looking remarkably similar to the modern cuticle and retaining a fine cast of the cellular structure beneath. None of the cutin structures survives intact in this deceptively familiar skin. Neither do they break down to the extent of becoming unrecognizable. A small proportion is formed of a series of closely related molecules called cutans. These, as far as we know, can remain intact indefinitely, or at least until the very slow cycles of reworking of the Earth's crust carry them down into the molten levels below. The DNA from Edward Golenberg's Miocene *Magnolia* leaf may have been an illusion, but for some of the lipids in the leaf's cuticle the Miocene epoch around 20 million years back was merely a transitory episode of their molecular youth.

plant skeletons?

Plants do not have mineral backbones like our own skeleton. For that kind of support they rely on organic molecules such as cellulose and lignin. They are, however, not without their own mineral inclusions, and these can linger on long after the cells around them have decayed. One of the most plentiful minerals within plant tissue is silica, absorbed in solution by the roots and then distributed around the plant's cells. These tiny particles of plant silica can survive in the ash of a fire, or as a residue of decayed plant tissue. They are not easily discernible from routine excavation of layers that appear relatively uniform to the naked eye. However, if small pieces of the excavated profile are set in resin and examined under the microscope, the layers of plant silica within ancient stems and leaves come into view. In the ancient settlement mounds of the Near East, they are especially plentiful. What seems at first sight like a uniform heap of collapsed mud-brick may prove, under the microscope, to be dominated by the ashy remains of plant silica.

These particles of silica, known as 'phytoliths', meaning 'plant stones', are in archaeological terms the plant parallel to the apatite in bones, the major mineral produced by the living body and the most durable element of it. Phytoliths are sometimes truly skeletal in that they have a strengthening function, sharpening the edges and honing the barbs along the edges of our more vicious grasses. At other times they are simply deposits of surplus dissolved silica, drawn into the plant from the soil water. They can survive longer than any other plant tissue. In Palaeolithic sites they may be the only plant fragments that persist. The Amud Cave in Israel is the latest Neanderthal site in the Levant, and is dated to 50,000 years ago. The cave sediments contain marked concentrates of grass phytoliths, with a significant number coming from the seed head. This may prove to be the earliest evidence of the collecting of grass seed that foreshadowed the development of cereal exploitation in south-west Asia several thousand years later.

Much can be learnt from the shapes of phytoliths. They are sometimes characteristic of particular plant genera, such as rice or maize. Sometimes they characterize particular environments, irrigated and dry-farmed crops differing in their range of phytolith types. Not much has been done on chemical variations within the silica particles, but in due course these variations may have much to reveal. Silica is not the only mineral residue from plant cells. Many crystalline and semi-crystalline substances deposited within the cell retain a characteristic form, and are sometimes extremely durable after death. One example is the layer of phytine crystals within a seed that acts as a trace element bank for the seed's future germination. Even after the seed has decayed, the microscopic phytine crystals can linger for thousands of years, free within an active soil.

This brief molecular survey has taken us further and further away from the genetic molecule at the heart of life, first to the proteins that preserve much of the DNA's sequence data, then to carbohydrates and lipids that are several stages removed, and then to biomineral molecules at the interface between the organic and the inorganic. While this progressive movement away from the genetic code diffuses some of the information content of the molecules, it far from erases it. Indeed, it sometimes reaches into a new area of information, passed over by traditional archaeology and DNA alike. Over and above this, extending the range of our molecular survey opens the way to study molecules with powers of persistence thousands of times greater than the persistence of DNA, retrievable both from more ancient deposits and from far more exposed locations in the archaeological record. We can now return to the archaeological record to see what information they can provide. A good way to start is by returning to one of the ancient farming journeys and looking beyond the DNA of the farmers and the crops they took with them.

beyond the maize field

Some time after wild teosinte was transformed to become domesticated maize, its cross-continental journey began. Maize cultivation spread from Central America, both to the north and to the south. Its path can be charted in both directions by finds of ancient maize cobs and maize seeds, often in the caves and rock shelters used by its cultivators. When the conditions of molecular preservation are right, this journey north and south can be related to the DNA both of the farmers themselves and of the maize cobs they left behind. In each case the DNA survives only selectively, within the heart of a seed or a bone that charted a specific course through the hazards of temperature, humidity, decay and fragmentation. But other biomolecules can brave these hazards with far less protection, carrying information that was in any case hard to glean from the DNA.

In the first few centuries after its domestication, the farming of maize had reached to the south of Panama. Twenty kilometres inland from Parita Bay in central Panama in the heart of the humid American tropics, a rock shelter called 'Aguadulce' ('freshwater') can be found. Excavations have demonstrated that this shelter was visited by humans 9,000 years ago and intermittently over the following five millennia. In the growing accumulation of debris they left behind there were grinding stones, perhaps for the milling of some kind of flour, possibly maize. This Panamanian rock shelter lies immediately south of the area in which maize agriculture began. Aguadulce is situated 2,500 km to the south-east of the stands of wild maize and climbing bean presumed to have been at the heart of agricultural origins in the New

World. It is on a route into South America along which the spread of maize and the other Central American crops can be traced, just as they can be traced to the north. The finding of grinding stones provides a convenient adjunct to that story of a spread of grain crop agriculture sweeping north and south, and laying the foundations for the early civilizations of the New World. But is that story skewed by an over-emphasis on the finds of maize? Dolores Piperno and her colleague Irene Holtz tried to find out.

They did so by examining fine cracks in the surfaces of the grinding tools. Working with a fine needle, they managed to prise out tiny particles of sub-cellular material, trapped within minute crevices in the surface of the stone. Some of these were composed of silica phytoliths, others were granules of starch. Both the silica and the starch particles retained some of their original shape and structure. On grinding tools that were 7,000 years old the characteristic phytoliths of maize could be made out under a microscope. This is an extremely early date for maize, and perhaps a direct marker of that first southward movement of maize farming, as well as confirmation that the production of maize flour was the purpose of the grinding stone. However, the maize phytoliths were not all that was found. Piperno and Holtz also recovered silica particles that came from two of the cucurbits that were so important to early American farmers – bottle gourd and squash – as well as from the lesser known tuber, lleren.

Higher up the sequence, in deposits laid down 1,000–2,000 years later, another series of grinding stones had entrapped starch granules as well as silica particles. From their size and shape, maize could still be recognized as one of the seeds being ground, alongside arrowroot, manioc and some type of legume. When the maize arrived at Aguadulce cave it was certainly ground up, but cave dwellers did not lose sight of a whole series of other seeds, roots and tubers that served as sources of flour.

As grinding stones from other sites were examined in a similar way, this picture of dietary range was placed in the context of time. At a shell mound on the coast at Monagrillo, maize was eaten alongside manioc and palm root between 3,000 and 5,000 years ago. In the village of La Mula, 2,000–3,000 years ago, the list had slimmed down to maize and manioc alone. Around 1,300 years ago, at the village of

Cerro Juan Diaz, grinding stones were found with traces of maize alone. The genetics of maize and its DNA variations had provided a framework for charting the spread of maize from its area of origin. Now these ancient grinding stones, and more particularly the tiny particles of silica and starch trapped within their surfaces, were charting the actual manner in which maize was incorporated into the lives and the food chains of people along the path of that spread. The new wonder crop did not spread like a flood, submerging everything in its path. It was gradually assimilated by people who were already grinding up a range of seeds, roots and tubers – plant foods they had known and depended upon for generations. Thousands of years were to elapse before the food chain had narrowed down to the kind of grain crop dominance that came to be associated with the monumental civilizations.

Roots and tubers are among the more elusive foodstuffs in the archaeological records. Seed crops and bones are a lot easier to track down, because of the visible and recognizable fragments they leave behind. We have, for this reason, tended to build our accounts of agricultural beginnings around them, and this has been reflected in the emphasis of DNA research. It is also reflected in a geographical bias in our accounts towards the kinds of seasonal environments where seed plants, open land and grazing herds are prominent features, regions such as the Fertile Crescent of south-west Asia. Our accounts have had a strong bias away from the tropical regions in which this emphasis upon seed foods is not traditionally as great. Here, roots and tubers have been much more prominent in the human food chain in recent history, and they do not persist within the archaeological record in the same ways as seeds. They will mostly be fragmented and disaggregated beyond recognition. Much of the unrecognizable plant tissue within archaeological deposits may be derived from these sources, and Jon Hather at London University has been leading the way in trying to make sense of it. Not many places in the tropics provide the kind of grinding stones for analysis that Piperno was able to examine, their fine fissures so conveniently trapping sub-cellular plant particles. More widespread are the chipped stone tools that are, less fortunately, fashioned to acquire smooth surfaces and sharp edges. But even these smooth surfaces can attract the attachment of plant-based substances

– substances that adhere for thousands of years, retaining information about the human past retrievable in no other way.

back beyond agriculture

On Buka Island, which lies towards the New Guinea end of the Solomon Island group, Matthew Spriggs and his colleagues from the Australian National University were excavating in a cave called Kilu. They dug down through just over two metres of cave deposit, first of all through scatters of pottery recognizable as just a few centuries old, to much earlier levels containing chipped stone tools rather than pottery. Throughout the deposits, remains of marine shells, fish, mammal and reptile bones were visible. It was clear what foods had been eaten, but the visible plant remains did not penetrate much below the uppermost levels. Further down in the excavation, the deposits and the chipped stone tools with them went back a considerable length of time, over 28,000 years.

Spriggs placed some of the small stone flakes under a microscope lens. He saw something he thought might be starch grains, not just on the younger material, but even on the oldest flakes recovered from the excavation. Staining with dyes specific to starch, and scrutiny under cross-polarized lights confirmed his suspicion. He took one basalt flake from the oldest levels in the cave and prepared it for scanning electron microscopy. Starch granules were clearly visible on the smooth basalt surface. Not only did the stone flakes have starch adhering to them, but long needles of crystalline calcium oxalate were also visible. These needles, or 'raphides' to use the technical term, are abundant in a family of plants called the Araceae. Their presence, together with the size of the starch granules, enabled Spriggs and his colleagues to argue that the flakes had been in contact with the tubers of taro. In modern times, taro is a staple crop of the islands, its tubers being made into puddings and bread, and fermented to produce a food called poi. Taro is not part of the island's original flora. The tubers may have been brought by sea, before planting, harvesting and scraping with stone tools, followed by cooking to remove the acrid calcium oxalate crystals still visible on the scraping tools. All this was taking place several thousand years before the earliest evidence of wild cereal collection in any part of the world.

The member of the team least surprised by the adherence and persistence of biological materials on stone tools would have been Matthew Spriggs's ANU colleague, Thomas Loy. By the time they worked together on the Kilu Cave material, he had spent a decade developing one of the more controversial strands in biomolecular archaeology. He had evidence that a lot more than starch granules adhered to stone tools.

blood from a stone

A 1983 article in *Science* carried a quite remarkable claim. For over a century, stone tools from prehistoric sites around the world had been scrutinized, measured and drawn, leaving us with pages of fine illustrations of the pristine surfaces of fractured stone. The usage of these tools was in large part a matter for speculation. Now a series of 104 such tools had been closely examined and, of these, all but fourteen had remains of their use adhering to them. Those remains were of hair, feathers, tissues and blood. Two years previously Jerold Lowenstein had published his findings of albumin proteins within fossil tissue, but he was working with protein molecules naturally encased and imprisoned while their host was still alive. Thomas Loy's traces of blood were quite different. They had remained exposed on the surface of artefacts with which they had made only brief contact, thousands of years earlier. The first examples he looked at were relatively young – up to 6,000 years old – and collected from four sites on the coast and in forest regions of Canada. Since then, he has gone on to seek out much older tools, inspecting their surfaces for any trace that could derive from ancient blood. He has had many positive results, and has claimed to be able to identify bloodstains left when a hunter struck his prey 200,000 years or more back in time. In these blood residues he sensed that a whole new world had opened up for prehistoric research. The most eroded parts of the prehistoric record could now be brought to light through molecular analysis.

The standard techniques of immunology and staining supported the argument that these were in fact blood proteins, but Loy wanted to go further than that and establish the identity of the blood. For this he made use of an elegantly simple property of the proteins, namely the manner in which they formed crystals. Many blood proteins display slight variations in their amino acid sequences, both within and between species. That is why Cavalli-Sforza could draw on them for genetic information. Those same sequence variations also affected the way in which the blood proteins crystallized. Slightly different amino acid sequences generated protein crystals with different shapes. To identify the shape and thus the identity of the ancient blood proteins on his archaeological finds, Loy re-dissolved a small part of the blood-stain from each stone tool under controlled conditions. He transferred the solution so produced to a microscope slide and allowed the blood to re-crystallize. With the minimal quantities used these crystals were only a hundredth of a millimetre in length, but this was large enough to display the kind of variation he was looking for. It enabled him to determine which animal species had come to grief at the hands of the prehistoric Canadians who used the tools. Among their prey were caribou, black-tailed deer, stone mountain sheep, moose, grizzly bear, snowshoe rabbit and sea lion. There was also some human blood.

In several more recent instances the archaeological record has surprised us with the survival of its complex organic materials, but Loy's blood residues on stone tools, as a form of molecular survival, have been more of a challenge to credibility. His results were greeted with not a little surprise and scepticism. The considerable persistence of proteins in intimate association with the mineral fractions of bone or shell was well attested – but on the exposed and smooth surface of a stone tool? It was hard to imagine, but there were two reasons to look at it seriously. First, there was clearly something on the surface of the tools. This 'something' was smeared across them in a manner suggestive of usage residue. That much could be established just by inspecting the tools by eye. Second, surfaces are remarkable things, and the chemistry at the interface between two surfaces may be quite different from the chemistry of the same substances when separate. This is something we have already seen with proteins clinging to the hydroxy-apatite crystals within bone. Perhaps these stone surfaces provided a similar raft to carry these large organic molecules through the millennia. Blood within a dead body may have no difficulty breaking down, but that does not mean that blood dried on the surface of particular minerals will act in the same way. We now realize that much of the biomolecular survival in bone is to do with the intimate contact between protein and mineral, rather than any intrinsic durability of the protein itself. Perhaps a similar thing might be true of blood on a stone.

It is probably fair to say that the key word remains 'perhaps' – we do not yet have a definitive model for this very exposed form of survival. Loy himself has suggested two mechanisms: first, that one group of proteins in the blood – the serum proteins – will, on drying, denature and desiccate to form a matrix which itself protects the coating from attack; and second, that clay minerals, well known to be strong surface actors, bind with the coating as it dries, and contribute to its longevity. Others have suggested that bonding with fatty acids in the soil could be an additional mechanism. If these issues can be sorted out, the potential would be great, as is well illustrated by a piece of work Loy undertook at the Toad River at the southern end of the Yukon Territory in north-west Canada.

Loy started out with a series of blades, fashioned from a stone called chert, collected from the region, but lacking a secure date or context. It was not clear how old they were, how they had been used, or what they were used on. Very many collections of stone tools that survive from the world's prehistory suffer this rather mute status. Loy, however, could detect sufficient traces on their surfaces to recapture some of the living world in which they were used. He found traces of conifer wood at the ends of some of the blades. These blades were apparently mounted on to a wooden handle as knives. He also detected repeated smears of blood, overlying each other. His crystallization methods narrowed these down to bison blood. Along with the blood were traces of hair and other tissue, whose presence and orientation on the blade suggested it had been used to cut through hide in the early stages of butchering the animal. There was even enough blood to acquire the fifty micrograms necessary for radiocarbon assay, as a result of which the blades were dated to the first millennium BC, situating them in the context of what was known about the changing environment. It had been assumed that in the first millennium BC, in this region, hunters in small groups had sought their prey from animals dispersed through the woodland. From scrutinizing the residues, Loy could assemble a different picture. The residues indicated that they also gathered together in larger numbers, on a seasonal basis, to exploit herds of bison in more open terrain.

potted histories

If this much adheres to the surface of a stone tool, how much more will attach itself to the rough and porous interior of a ceramic pot? Fired clay vessels have only been a major feature of human life over the last 10,000 years, roughly corresponding to the epoch of agriculture. In that most recent episode of the human story, they have become one of the most familiar archaeological materials alongside the much more ancient corpus of stone tools. As with stone artefacts, traditional approaches to analysis involve describing size and form and ordering into stylistic sequences, rather than establishing what was carried within them. This is understandable, considering that pots are generally recovered as fragments of empty vessels – particularly empty after an army of archaeological pot-washers has scrubbed them clean. At a molecular level, they are not as empty as they at first appear. Because of their porosity and lack of a glaze, they can have an enormous surface area, mostly contained within minuscule pores. A large unglazed storage jar could quite conceivably have a surface area the size of a football pitch.

Since the 1970s, archaeological scientists have attempted to recover any lingering residues in order to work out how the pots were used and what they had contained. The first attempts involved rinsing the pot fragments with organic solvents and then analysing the resultant solution through infra-red spectroscopy, by then a well established method of identifying organic molecules in very small quantities. With this method it was possible to establish that lipids were preserved, and, for example, to speculate on the various oils, pastes and preserves that had been transported across the Roman Empire. Their lipid residues could be sought within the large, pointy-bottomed amphorae that are a familiar find within Roman shipwrecks, and as prestigious items in high-status graves along the Empire's periphery.

By the 1990s, the range of spectrographic methods available had diversified, and Richard Evershed at Bristol had added several refinements. He heated his solutions to high temperatures in order to vaporize any molecular residues. These vaporized molecules could then be sorted by using a combination of gas chromatography (GC) and mass spectrometry (MS), predictably referred to as GC/MS. In general terms, GC/MS can be likened to a molecular racetrack, which ends, not at a neat finishing line, but at a crash barrier. The molecules all start off at the same starting line, but as they get going round the track they separate into a rank order according to variations in structure and mass. That is the GC element. The MS element is the crash barrier. The different molecules are forced to swerve, under the powerful force of a strong magnetic or electric field, and hit the barrier in different places, again according to variations in mass and other structural properties. From the order the molecules move around the track, and the point they hit the barrier, Evershed could work out a great deal about their structure. Furthermore, he could look for a whole series of different molecules by scrutinizing his crash barrier after a single spin of his sample round the track.

After a number of trials with this highly sensitive method, Evershed realized that chemical residues were not at all uncommon. Something like half the pots recovered from archaeological sites retain some trace of what was inside them. GC/MS methods could describe those chemical residues with remarkable precision. Evershed was keen to push the method as far as it would go. An important point in his favour was the fact that most prehistoric and mediaeval pottery had not been glazed. Such smooth surfaces as existed on ceramic vessels in prehistory and the Roman period were generally achieved through burnishing the still damp clay with a stone, or coating it with a fine clay wash or 'slip'. Furthermore, the clays were tempered with some non-clay mixers, which could be limestone, dung or some other organic matter, and which enabled them to be fired at relatively low temperatures. Thus the prevalent pottery types, even up to the Industrial Revolution, had a relatively coarse and porous fabric. A second point to consider was that their production, and very often their use, involved exposing them to great heat. Although organic molecules survived, they would be transformed by the heat.

Bearing these points in mind, it seemed likely that the most useful group of biomolecules to study was the lipid group. By definition they are insoluble in water and so would not get washed in and out of the pot's pores. In addition, many were reasonably tolerant of heat. Unglazed pottery can act as a kind of sponge to groundwater after a while, and there will be a fair amount of movement of water through buried sherds of pottery. The lipids are tucked away within the microscopic pores of the pot and, being insoluble in water, stay put, however much the water moves back and forth. Moreover, the presence of water does not encourage breakdown, as it does in the case of the long-chain molecules. The side branches of lipid molecules may be lost or transformed, but the original core of the molecule may stay intact for a remarkable period.

Before long, Evershed turned his attention to another of the key food sources habitually overlooked by prehistorians because of the lack of any obvious trace in the archaeological record – the green vegetable foods. Green tissue, in the form of leaves, buds, stems and flower heads, plays a key role in vegetable diets and is particularly important in relation to vitamins and minerals. Yet we archaeologists know very little about the early use of these foodstuffs. We know virtually nothing about that important group of green vegetables, the brassicas, which includes broccoli, cabbage, cauliflower, kale and kohlrabi, simply because these leafy green tissues are too fragile to survive intact.

Evershed was aware that, while those large soft expanses of green tissue were about the most vulnerable of all tissues to rapid breakdown, there were certain chemical elements within them that were quite the opposite. The waterproofing molecules that prevent those soft tissues from drying out, such as the cutins and waxes, are among the most obstinate molecules of all. They can persist, in slightly modified form, for a seemingly indefinite period.

Ordinary cabbage, one of the brassicas, owes its glossy protection to a long-chain molecule, composed of hydrogen and carbon and called 'nonacosane', together with a few very close relatives that have a little oxygen within them as well. That little family of molecules hit

Evershed's GC/MS crash barrier fairly close to one another, providing him with a characteristic read-out for the group. He embarked upon a search for the familiar cluster in a group of mediaeval potsherds from West Cotton, a rural settlement in central England, all from unglazed pots used in everyday life. A high proportion of them produced a variety of volatile lipids, separating out within the gas chromatograph to produce a range of peaks. A cluster of those peaks repeatedly matched the nonacosane cluster, and in five of the potsherds they were the only peaks. Those pots had a lipid signature remarkably similar to the cabbage leaf itself. They must have been cooking pots for brassicas, used time and time again for the preparation of those particular foods. The leafy vegetables that completely elude the conventional archaeological record were now showing up, not as an incidental food, but as vegetables so important that some pots were used specifically for them.

a rich and diverse feast

By looking at a range of molecular markers, we can place the spread of major world crops in the context of the considerable diversity of prehistoric diets that conventional archaeology has failed to detect. There are many ways of living in and using the natural environment, exploiting countless plants and animals – tubers, roots and leaves as well as meat and grain. The spotlight has also shifted to other parts of the global stage. It has cast its beam on the world's vast tropical regions, poorly visible to conventional archaeological methods, but now progressively coming into focus, their prehistory just as rich and complex as that of the better known temperate and sub-temperate zones. These same molecular methods allow us to probe further into what early people were actually doing with these foodstuffs, how they were taking fresh plant and animal tissue and forming it into the components of a meal. Take, for example, those cabbage pots that retained the molecular residue of the cabbage leaves' waxy cuticles. Evershed's analyses went a step further than simply identifying the food within the pot. The GC/MS methods he employed provide an extremely sensitive route to targeting scarce remains of lipids at particular points, and subjecting them to analysis. Using this technique, he took one of his ancient cabbage pots and searched for the cuticle waxes at different points in the interior. Inside the pot at its base the waxes were present, but at a fairly low level. He took samples again at different levels up the side of the interior. At a certain level, the lipid signal peaked. He had found evidence of boiling, and of the level which the boiling water reached and at which lipids tended to float.

More evidence of boiling came from other mediaeval coarse pots. These were empty pots blackened by fire on the outside, but solvents applied to the inside picked up a lipid residue. Evershed's GC/MS analysis was once again targeted upon several points within the pot, from the top, the side and the bottom. This time he found that boiling was not the only aspect of the pot's history to leave a lipid trace. First, the higher within the vessels they tested, the more lipids they recovered. This, he suggested, came about through boiling water-based meals within the pot, causing fats and other lipids to rise to the surface. Second, the smaller quantities of lipids at the base were of a quite different character from the larger quantities at the top. The ones at the base matched with beeswax, while the lipids higher up matched with animal fats. The beeswax was clearly not associated with the bubbling stew, but was sufficiently well bonded with the pot surface to remain in place throughout the cooking process. Around the world it is still possible to find potters who take a newly fired pot, still warm, and smear the inside with beeswax before anything else has had a chance to seep in. This initial proofing of the mediaeval pots from the West Cotton village improved their performance, as they were used to prepare meat stews. Once again, looking at molecular patterns in space across the surface of an artefact allows us to construct a sequence through time.

The analysis of seeds and bones that had expanded in the preceding episode of archaeological science had allowed lists to be assembled of the visible foods of antiquity. Now molecular analyses were going further than simply adding to that list the tubers and vegetables that had hitherto remained invisible, and was moving towards the detection of how those foods were being treated and prepared for the meal. This was opening two further windows on the human past. The first relates to our physiology and the way our digestion works.

'External digestion' is a catch-all phrase for a range of things we do to get the breakdown of food started before it enters our body – such things as grinding, cooking and so on. It is a highly characteristic feature of our species, and is an important factor in the very diverse diet we adopt. We do not possess a particularly remarkable stomach, compared for example with the multi-chamber affair of a ruminant mammal, and yet our dietary range has spread to encompass some surprising challenges. The initial ingredients for our meals can include things too hard to chew or too intractable to digest, and things that are poisonous, purgative and inflammatory. If we relied on 'internal digestion' such a range would be impossible, and the use of grinding, fire and fermentation in external digestion has been a major feature of our ecology since long before agriculture began. The Pacific Island taro traces of 26,000 years ago and the similarly ancient evidence of grass-seed consumption in south-west Asia both indicate the antiquity of such transformations. Both taro and grass seed require careful treatment before consumption.

Work on ancient seeds and bones had given us a partial list of ingredients. Ancient biomolecules were both adding to those ingredients and shifting our attention from the ingredients to their preparation as a meal. There may be hundreds of ways in which any two or three ingredients can be transformed into a culinary delight and shared within small or large communities of people eating together. How those transformations are carried out tells us a great deal about past societies, the numbers of people who lived together, and the manner in which the passage of time was marked out by feasts.

Anthropologists have shown great interest in the meal. In an attempt to see some pattern in the great diversity of methods of food preparation Claude Lévi-Strauss devised a culinary triangle. At the points of the triangle were the 'raw', the 'cooked' and the 'rotten'. The various culinary practices found around the world could then be situated between these points. Patterns within the triangle, and further variations in preparation, notably between roasting and boiling, could in turn be related to social practice and ritual that defined relationships and shaped the world. We have only to think of the status of the roast joint of meat, carved by the head of house, a ritual focus mirrored in different social gatherings in many societies. Boiling can be picked up by lipid analysis, but what about roasting? Evershed examined lipids within sherds that had broken away from a characteristic type of large flat dish found on the same sites as the cabbage pots. The component fatty acids were assayed, and a potential source identified. The suite of fatty acids on the dishes derived from animals rather than plants, and from a quite distinct group of animal fats. They had dripped from a roasting pig. The lipid spectrum provided the link between some rather featureless pieces of broken pot and the spit-roasting of a suckling pig that provided an important social focus for a celebratory meal, its dripping fat caught below in these unglazed dishes.

The third point of Lévi-Strauss's culinary triangle is 'rotting'. In biological terms, what that means is some form of digestion by a relatively benign micro-organism. Such external digestion is fundamental to numerous food preparation processes, including the rising of bread, the brewing of beer, the fermentation of poi from taro, the production of yoghurt, bean curd, soy sauce and brown-leafed tea. Detecting it in ancient samples is not easy, for two reasons. First, these micro-organisms are particularly efficient at breaking down substrates to very small, and thus non-persistent, molecules. Second, their actions are similar to decay after disposal in the ground, often enacted by similar, if not the same, micro-organisms. However, a Cambridge research student, Delwen Samuel, managed to pick up traces of fermentation by bringing together some of the approaches above, and looking at starch granules within ancient pots.

In fact, she found a great deal more than just starch granules coated on the inner surface of her potsherds from ancient Egypt – enough material within the modest smear to instil a little anxiety into generations of archaeological pot-washers. What their scrubbing brushes had cleaned away was transformed, under the microscope, into a rich array of foodstuffs, among them recognizable fragments of cereal grain and other seeds. At a higher magnification the granules of starch came into view, their normally even surfaces pitted by tiny holes. She was observing the direct action of fermentation, with individual yeast cells having digested the surface of the starch granules. Following a range of microscopic and molecular analyses, these seemingly unremarkable residues within ancient pots provided enough information for the bread and beer of the ancient Egyptians to be reconstructed. After a few years of her research, a well-known brewery that was funding her work was ready to re-create the beer the Pharaohs sipped. A limited edition brew went on sale at Harrods in London at £50 a bottle, and speedily sold out.

sharpening the focus: the isotopic signature

For molecular archaeology to become a real possibility, the sample requirement had to shrink to within the range of what survives the ravages of time in an archaeological deposit. Like the DNA residues explored in earlier chapters, proteins and lipids can now be detected in quantities of less than a millionth, sometimes less than a billionth of a gramme. With methods of such sensitivity, individual cells, or individual points on the surface of an artefact, can be chemically studied. That sensitivity can take us one step further yet, to examine the atoms from which the ancient biomolecules are built.

But why should we want to look more closely at the atoms – is there anything further they can tell us about the past? In fact, there is not a great deal of atomic variety in the complex biomolecules featured here. For all their intricate diversity and variation, they are very largely built from the same five elements – carbon, hydrogen, oxygen, nitrogen and sulphur. But each element is not always present in a uniform state, and the different states of the constituent elements, or 'isotopes' as they are known, are themselves a mine of information. Carbon, for example, is an element that can occur as one of three isotopes, designated C12, C13, and C14. Those numbers are not arbitrary; they indicate the differences in mass of the atoms. An atom of the lightest isotope, C12, is twelve-thirteenths of the mass of an atom of C13, which in turn is thirteen-fourteenths of the mass of an atom of the heaviest isotope, C14. All three can become incorporated into precisely the same molecules. Generally, they will behave in a similar way, except that the heavier isotopes tend to be more sluggish in certain processes. This causes them to accumulate differentially, and that is when they can capture and retain information useful to archaeologists.

One way in which this sluggishness of heavy isotopes makes itself known is through the food chain. As carbon passes from the atmosphere into plants, on to herbivores, and from there to carnivores and top carnivores, much of it is burnt off as carbon dioxide from each step in the chain. More of the lightest isotope is released in this form, leaving more of the heavier isotopes to accumulate within the solid tissue. So the proportion of the light carbon isotope drops with each step through the food chain, and what gets incorporated into the proteins within an organism remains as a marker of where those organisms are placed in the food chain. In other words, it becomes a record of past diet. If we compare the collagen of three healthy people, one a vegetarian, one with a mixed diet, and one an avid meat-eater, then the proportion of heavy isotopes grows progressively as we move from vegetarian to avid meat-eater. In ancient bodies, we only use the ratio between C12 and C13 for dietary analysis. C14 is unstable, which renders it invaluable for one of our most important dating methods, but no good as a dietary signal.

The stable isotopes of other elements behave in a rather similar manner. Nitrogen is also common in biomolecules, particularly proteins, and occurs as one of two stable isotopes, N15 and N16. In a manner similar to carbon, the heavier isotope is progressively enriched when moving up through the food chain. Even within a single body, the relative fractions of the different isotopes will vary, for example between the bone, soft tissue, milk and any surviving strands of hair. The isotopic patterns of a range of elements come together as a dietary diary, a record of who ate what, and when, in the prehistoric past.

The virtue of looking at a variety of elements, rather than at any single one, is that each element is subject to other factors, and not just to position on the food chain. Sometimes these other factors are not of direct interest to archaeologists. For example, isotopes of nitrogen are very much affected by the soil micro-organisms that find themselves caught up in the human food chain. But on other occasions these additional factors add to the detail we can reveal about past human diets. Carbon isotopes are a case in point. Certain grasses, for example the cereal maize, carry a particular isotopic signature, which is preserved in the bone collagen of someone who consumes a lot of maize. This particular signature derives from the distinctive way in which many tropical and sub-tropical grasses, including maize, capture the sun's energy. They employ a distinctive photosynthetic pathway, which incorporates carbon isotopes in a very different ratio from most other plants. While the resulting C12/C13 ratio is fairly common in the tropics and sub-tropics, as maize spread north and south into the temperate zones the ratio became highly distinctive. People consuming a lot of that maize would incorporate the same unusual ratio in their own bones. In fact, maize provides one of the key examples of how isotopic information can reveal a great deal of information about the human past.

The general chronology of the spread of maize can be reassembled from finds of the cobs and grains. These show that it had arrived in the eastern United States by around 800 BC. This is the region in which the mound-building, metalworking Hopewell culture flourished from around 200 BC. It is an example of a 'complex society', the kind of society for which many have presumed cereal agriculture was a precondition, and the seeds provided evidence that it had been around for a few centuries before social complexity grew. However, possessing maize cobs is a very different thing from depending upon them. As Piperno's study of the surface of ancient grinding stones has shown, there was more to the early farming food chain than maize alone. In the 1970s, Nick van der Merwe tested the notion of maize dependency by looking at the collagen isotopes from skeletons excavated from sites in the eastern United States. In skeletons of 5,000 years old, he recorded a carbon isotope signature typical of temperate zone foods, which is what we would expect, well before the appearance of maize in the region. When he moved to younger skeletons concurrent with the mound-builders, to many archaeologists' surprise he found more or less the same signal. There was no sign of maize playing a major dietary role, even after it had been in the region for several centuries. It was not until AD 800, long after the mound-building Hopewell culture had waned, and a millennium and a half after the first maize in the region, that the C13 levels show a steep rise. In the region, this was the earliest sure sign of maize consumption on a large scale. Rather than stimulating and initiating the mound-building culture, the consumption of maize as a staple crop had actually followed it. What is more, the shift from maize as a dietary supplement to a major agricultural source did not seem to be particularly beneficial. In fact, the state of the bones themselves indicated the later peoples to be suffering from the narrowing of their diet from its greater diversity in Hopewell times.

dissecting the diet

Isotopic analyses can be employed to distinguish between herbivores and carnivores, between land-based and sea-based diets, and to pick up certain foodstuffs such as maize and the legumes. It has now become possible to go even further and to break down this dietary diary according to finer distinctions within the food chain. Let us return to Richard Evershed's analyses of ancient lipids. He took advantage of a refinement in the GC/MS technique which allowed it not only to identify minute traces of ancient molecules but also to record their isotopic make-up. The dripping dishes that caught the fats from the spit-roast suckling pig provide an example of what this extra information can tell us. This refined technique allowed Evershed to measure the isotopic balance of the specific carbon atoms locked up in fats collecting at the bottom of those mediaeval roasting dishes. Pig stomachs fractionate carbon in a manner quite distinct from the ruminant stomachs of other domesticated herbivores. The distinctive pig signature matched perfectly the signature from the fats recovered from the dish, a useful corroboration of his interpretation. From other ceramic vessels he was able to go further and distinguish another animal product.

Dairying plays a major part in the animal husbandry of the western world. In some cases it has lain at the centre of agricultural life, with milk, curd, cheese or yoghurt as the prime reasons for keeping animals in the first place. This, however, is not a universal practice. Many people in the world have difficulty in digesting milk. Moreover, it is not clear that dairying has anything like the antiquity of agriculture. Animals may have been domesticated for thousands of years before the practice of diverting milk from the animals' young to the human food chain was introduced. There again our archaeological evidence has been shaky, dependent on the presence of artefacts tentatively identified as cheese-strainers, for example. It is an important and contentious issue in prehistoric agriculture. Some would argue that dairying was integral to farming from the start, and others would argue that it was a more recent, geographically restricted, phenomenon, resulting from a separate and distinct revolution in human ecology. While the artefacts were rather inconclusive on the issue, Evershed put his mind to discovering whether molecular evidence might resolve it.

Distinguishing meat from milk from the same animal is not a straightforward task. The problem with many molecules found in milk is that they are also found in other tissues, including meat, and so cannot serve to separate the two. There are nevertheless some lipids that are much commoner in milk than in meat. They are distinguished by their relatively high proportion of short-length fatty acids. Evershed attempted to seek these out, but found that the shorter fatty acid molecules were significantly more likely to break down than their longer counterparts, badly skewing any ratio between the two in ancient material. At this stage, Evershed turned his attention to the isotopes, and found the differentiation he was after. If the same lipids are compared in milk and meat from the same animal, the milk lipids are distinguished from the meat lipids by a significantly lower proportion of the heavier carbon isotope. Again, by using GC/MS to target particular fatty acid molecules, and then assessing their carbon isotope ratio, Evershed found traces of milk in some of his mediaeval pots. He is now pushing the analyses well back in time, to establish conclusively when the practice of dairying began.

a dietary diary

The milk residue in an ancient pot may represent as little as a single meal. The collagen within our bones accumulates through life and records the average diet over a decade or so. We can extend the search to other parts of the archaeological record, for organic remains that record diet on other time scales. One such residue is hair.

Whenever yuletide excess leads on to a January diet, then it should show up in my beard trimmings. Soon after Christmas they will be laden with heavy isotopes, as a succession of meat-rich meals work their way through my system. By January, a repentant intake of light salads will have significantly changed my beard's isotopic composition. This will not be discernible in my bones. They will have merged the dietary signal of several years' consumption, but our hair monitors short-term changes. A thread of shoulder-length or longer hair is like a dietary diary, with seasonal changes in nutrition recorded.

Hair is composed of a tightly wound protein called keratin, the same substance of which wool, horn and feathers are made. A sizeable proportion of the keratin molecule is composed of an amino acid called cystine, which forms particularly strong bonds that endow keratin both with its durability and with resistance to decay. It does not take much to inhibit the fungi that normally decay hair. It may be the poisonous metal seeping from a lead coffin, or more familiar inhibitors, such as extremes of wet and dry. Ancient hairs can be found both free within an archaeological sediment, or still rooted to an ancient body. Some of the best preserved bodies that have featured in various chapters of this book have intact hair. Steve Macko from Virginia University decided these would provide a sharp and direct record of particular diets in the past.

One opportunity for study came from an ancient traveller, frozen solid 5,300 years ago in the Tyrolean Alps. This 'Ice Man', or 'Oetzi' as he is locally nicknamed, set off high in the mountains, equipped with axe, bow and a sheath of arrows. Before hypothermia felled him, his eyes were presumably peeled for an ibex or a chamois – those mountain grazers, fragments of whose fur he used in his clothing. Some time after death, his hair became detached from his body and lay in tufts, caught up in his clothing. Macko took a few strands of that hair and tested them for their isotopic signature. He anticipated a fairly meaty signature, as everything about the man and his equipment suggested a hunter. Instead, the isotopes presented him with a surprising result.

At the time of his frosty death, Oetzi's hair displayed the isotopic profile found today not among hunters but among their complete antithesis, vegans. How could that have come about? What was a fur-clad vegan doing with bow and arrows high in the Tyrolean Alps? It was not a question that could be answered by the hair isotopes alone. Macko was only one of several researchers applying the new biomolecular methods to Oetzi's body and paraphernalia. Konrad Spindler, the archaeologist in overall charge, had read about the remarkable findings of an Australian researcher on the surface of stone tools. He wrote a short letter to Tom Loy, inviting him to come across to Innsbruck to inspect the finds.

Oetzi's toolkit provided Loy's keen eye with a veritable feast of data and, yes, there was ample evidence the ancient man was a hunter. The bows, the arrows, and in addition the copper axe and stone knife all had the traces of blood Loy had come to recognize on stone implements from around the world. On the arrows, the traces of blood stretched back 30 cm along the shaft. These were big animals at the end of Oetzi's line of sight. There were also fragments of hair and feather, and sheets of collagen adhering to several of the tools, all contributing to the picture of hunting, skinning and bone-working. Loy's data seemed to clash with Macko's. Yet there were some points of commonality.

Adhering to the copper axe-head, especially around the haft and under the thong, were starch grains, the kind of trace that Piperno had encountered on Central American grinding equipment. From the manner in which they were distributed, Loy was able to speculate that Oetzi was rebinding the thong around his axe-head as he ate his last meal, a starchy meal of plant foods. Oetzi may well have been a seasoned and experienced hunter, his consumption of meat showing up in the isotopic pattern of his bone collagen. The strands of hair do not describe that life, but record a more transitory state, some time before his death, when he was consuming just plant foods, and perhaps lamenting his failure to bring down the prey he craved. We now know from fragments of meat fibre in his intestine that he eventually succeeded. It may indeed have been his desire for meat that took him to the perilous heights from which he failed to return.

9

friends and relations

individuals

There was always a hope, ever since it became clear that ancient biomolecules survived, that they might lead us to rediscover some of prehistory's individuals. We might find out something about them, who their families were, and with whom they made contact for the exchange of goods and ideas, for the establishment of social connections, and for marriage partners. This was in part because of the particular powers of DNA, the molecule that was attracting the most attention, and in part because of the possibility of working with very tiny traces indeed. It was, at least in theory, feasible to trace an individual's path through life by typing the DNA from shed hairs or flecks of skin. Even a fingerprint might leave a mark of its maker behind. Just as molecular archaeology was gaining momentum, so those same possibilities were revolutionizing forensic science. Individuals were indeed being identified from minimal traces at the crime scene – a speck of blood, a semen stain, or even a discarded cigarette butt that had been in contact with the wrongdoer's lips. From these tiny traces, the polymerase chain reaction was able to bulk up sufficient amplified DNA for it to be typed and linked to a specific individual.

A method for tracking down individuals from their DNA was already in place before PCR came on the scene. It is what has become known as 'genetic fingerprinting'. What happens is that selected restriction enzymes are used to break up the entire DNA blueprint into strands of different lengths, according to where the enzymes bind. That multitude of created strands is then allowed to migrate along an electrophoresis gel to generate a characteristic sequence of bands, the 'fingerprint'. As most of the genome is identical in most of us, the great majority of bands will also be identical. However, the various hot-spots along the genome will lead to strands of varying lengths and thus minute differences will be encountered between individuals, according to how distantly or closely related they are. The entire fingerprint is unique to the individual concerned.

This remarkable method was unfortunately not transferable to the archaeological record, for one simple reason. As soon as the cell dies and the active DNA repair mechanisms cease, the DNA itself begins to fragment. In a period of a few years it becomes broken up into the very short lengths that are the stuff of ancient DNA science. The restriction enzymes used in fingerprinting will only add to a process of fragmentation that is already heavily advanced. As with all ancient DNA projects, we have to work on variation that is sufficiently localized to be retained within those fragmented strands.

In its original form, genetic fingerprinting was also limited in its forensic applications by the quantities needed, which were often in excess of what the criminal left at the scene of the crime. This constraint disappeared with the advent of PCR, opening the way to the acquisition of a fingerprint from minute traces. A second modification of the method opened the door to identifying past individuals from traces of their ancient DNA. This entailed a shift of attention from the whole corpus of DNA within the cell, to short, informative strands of DNA instead. It was an approach that Erika Hagelberg took with some considerable success.

The opportunity to develop such an approach arose when she turned her attention to two criminal cases that were very much more recent than traditional archaeological remains, but not so fresh that modern DNA methods could be applied. One of these cases involved a murder victim, a fifteen-year-old girl whose decomposed remains were unearthed eight years after her death in 1981. The other involved a man who had died in a swimming accident two years previously. It was suspected that he might have been involved with murder, but as perpetrator rather than victim. It was thought that his true identity was the notorious 'Angel of Death' of Auschwitz, Dr Joseph Mengele. In both these cases there was an uncertainty about identity, and DNA science was brought in to resolve that uncertainty. However, even though the bones examined were from bodies that had only been dead for a few years, the human DNA within them proved to be heavily fragmented, apparently more so than some of the better conserved prehistoric bodies. Although the DNA under examination was 'younger' than some of the DNA still active within the bodies of the scientists doing the work, in experimental terms it was 'ancient'.

The approach that Hagelberg took was to search for the most highly variable hot-spots she could find within the genome. They would need to be sufficiently compact to show up within the short fragments to which their DNA had been reduced. Furthermore, they would need to be even 'hotter', in evolutionary terms, than the regions used to explore migrations and agriculture, as here Hagelberg was looking at specific individuals rather than broader lineages. The obvious choice was a feature known as a 'microsatellite'. This is a short strand of highly repetitive DNA, whose length is extremely variable. The repetition is often composed of just two or three bases, repeated in tandem any number of times. When the repeat unit gets larger, and the overall length goes much beyond 150 base-pairs, the term used is 'minisatellite', but the principle is the same. In both the studies, Erika Hagelberg turned her attention to microsatellites made up of variable repeats of a cytosine–adenine pair. Such microsatellites are found on a number of different human chromosomes, and their variation might allow a kind of fingerprint that would hold true even in fragmented DNA.

In each of the two cases, she needed access to tissue from close living relatives. The bereaved parents of the girl thought to be a murder victim provided blood samples, as did Joseph Mengele's wife and son. From the samples, Hagelberg and her colleagues built up a picture of microsatellite variation among two parents and their offspring in each of the two suspected family groups. If either identification were correct, then the range of microsatellite variation in the offspring should be neatly attributable to an even inheritance from the two parents. In both family groups this was found to be the case. In England, the bereaved parents were now able to bury their daughter, and on the other side of the Atlantic in South America, Auschwitz's Angel of Death had been identified.

This approach has generated a rich new strand of forensic science.

There are, tragically, a vast number of 'missing persons', unceremoniously dumped in the context of totalitarianism and genocide around the world. In shallow graves especially, the transformation from intact 'modern' to fragmented 'ancient' DNA takes only a few years and, for many, ancient DNA techniques may offer the only means of identifying lost relatives. Erika Hagelberg and others have continued with this forensic project, at the same time exploring extending the approach to the less harrowing identifications from the more distant past.

the death of the romanovs

One renowned massacre, on the boundaries of living memory, took place in 1918, in the Urals of central Russia at Ekaterinburg. What we know of events derives from a dossier compiled over the following year by one Nikolai Sokolov. His dossier records the events of the night of 16 July, or shortly thereafter. The last Tsar of the Russian Empire, Nicholas II, was taken down to a cellar of the house that was serving as his makeshift prison. Accompanying him was his Tsarina, their five children, their doctor and three servants. Down in the cellar, they were shot and mutilated. A plan to dispose of them down a mine-shaft went wrong, and the bodies ended up in a hastily dug pit. The backfilled hollow was driven over to flatten it, and drenched with sulphuric acid in an attempt to erase the evidence. Seventy-two years later, twenty miles outside Ekaterinburg, two amateur historians recovered nine badly damaged skeletons from a shallow grave.

A consortium of English and Russian scientists was invited to investigate the discovery. The problem facing the international team, which included Erika Hagelberg, was at least one step more complex than that of the more recent bodies of Mengele and the young murder victim. The treatment of the bodies was clearly unfavourable to molecular conservation, but that had been true of the younger remains as well. In all cases, the DNA would be heavily fragmented. An added difficulty in identifying the massacred bodies of the Romanovs was that the very purpose of the killing was to extinguish the family line. There were no spouses and offspring to provide samples. However, the Romanovs belonged to one of the best documented families in the world, and the possibility of tracking down distant surviving relatives remained open. As this would involve tracking through several generations, a different approach was required from that used with Mengele and the murder victim. At the remove of only one generation, nuclear microsatellites would remain sufficiently intact to link parent and child. In the context of tracking over several generations, those same patterns begin to become confused by the several episodes of sexual recombination of DNA sequences. This does not render them useless, and the research team did manage to establish from them that five of the skeletons conformed to the pattern expected from a family group. But to get a surer result, what they needed was to follow a 'haplotype' pattern that would remain much more intact between generations. For this they used the sequence that has underpinned so many ancient DNA projects, a hypervariable segment of the mitochondrial control region.

What they were after were control region haplotypes that would remain faithful across the different maternal lineages that converged on the Tsar's family. If these led to living relatives, then the identity of the bones from the shallow grave might be established. The two maternal lines they tracked were the ones running through the Tsar and his Tsarina respectively. The Tsar's mother was Dagmar, daughter of Louise of Hesse-Cassel. Her mitochondria would be borne by the Tsar, but not by his children, who would carry the Tsarina's mitochondria instead. The continuation of the Tsar's mitochondrial genome would only occur through his maternally linked female relatives. He had a sister and an aunt sharing his mitochondrial genome, and both these women had living descendants following the maternal line. The great-great-grandson of his aunt and the great-granddaughter of his sister are both alive today, and each agreed to provide blood samples.

As for the Tsarina, her mother was Princess Alice, the second daughter of Queen Victoria. The Tsarina had a sister, Princess Victoria of Hesse, and all these women naturally possessed the same mitochondrial genome. That genome was in turn shared by Princess Victoria's daughter, Alice of Battenburg, and her son Philip, who went on to marry another of Queen Victoria's descendants, the reigning British Queen, Elizabeth. Philip, now Duke of Edinburgh, was Chancellor of Cambridge University when Erika Hagelberg was working there. He too agreed to supply a blood sample.

Having carefully targeted the appropriate descendants, always following the maternal line, the team had managed to identify one mitochondrial haplotype that would match that of the Tsar, and another mitochondrial haplotype that would match that of the Tsarina and all her children. If Sokolov's account of the massacre was accurate, and the two amateur historians had actually recovered the bodies of the imperial family, then this would be the pattern: among the bones, those of one adult male would have carried the first of the two haplotypes, that of the Tsar, and those of one adult female, four younger females and one younger male would have carried the second, that of the Tsarina. We would also expect four other adult males (the servants and doctor) to match neither. That would constitute a direct translation of Sokolov's account into mitochondrial genetics. With that model result in mind, the team set to work recording the hypervariable sections of the control region, as preserved in the recently unearthed bones.

What they recovered was a striking, but incomplete, match. First of all, there were four adult skeletons matching neither haplotype – these could correspond to the three servants and the doctor. Second, the Duke of Edinburgh's haplotype matched that of the adult woman and of the younger skeletons, which numbered not five but three. It seemed that two of the children were absent from the shallow grave. The haplotype of the remaining adult male almost matched that of the living descendants of Louise of Hesse-Cassel and, by inference, the Tsar. At one point along the Tsar's sequence the nucleotide was mixed, a fascinating reminder that these DNA regions are studied because evolution along them is so rapid. In addition to providing compelling evidence that the Romanov family had been rediscovered, this slight difference within individuals separated by four generations has left a graphic biochemical trace of evolution in action.

breaking the link with the present

The Ekaterinburg bones demonstrated how, through ancient DNA analysis, the study of kin relations could be taken back in time to a poorly conserved archaeological deposit, albeit a rather recent one. It remained, however, a rather special case. We rarely have such an extensive family tree with which to reach back from the living descendants to the excavated bones of their presumed ancestors. That is rare enough in burials 100 years old, let alone those from conventional archaeological deposits.

One research group that from the outset has been tackling the possibility of approaching kin relationships through ancient DNA is the Göttingen laboratory of Susanne Hummel and Bernd Herrmann. They too have been attempting to make sense of archaeological burial groups by amplifying microsatellites and other fast evolving units. In 1993, archaeologists were able to provide the Göttingen team with a family group that could be analysed in as fine detail as the Romanov group. They found it in an ancient church in Lower Bavaria at Reichersdorf. Here eight skeletons had been excavated from beneath the chancel floor. Set into the surrounding chancel walls were the memorial stones that provided Julia Gerstenberger at the Göttingen lab with the template she needed. The stones recorded seven male members of an aristocratic line, the Earls of Königsfeld. That seemed close enough to look for a match with the eight burials. The first of the named earls was Hanns Christoph, who died in 1546. The last was Georg Josef, who died two centuries later in 1749. The remaining five were the intervening earls in the patrilineage. The church clearly served as a sepulchre for this male aristocratic line, and it seemed quite feasible that the excavated skeletons might be of the earls themselves. Even prior to the team testing this through ancient DNA analysis, there were hints from examination of the skeletons that the match would not be 100 per cent. At least one of the bodies had both the skeletal form, and fragments of the dress, of an adult woman, and there were no stones in the chancel commemorating her.

The Göttingen lab had also been using ancient DNA as a means of sexing ancient skeletons. There is a gene involved in tooth development, the amelogenin gene, which is carried by both the X and the Y chromosome. The gene is of a slightly different length in the two chromosomes. When an appropriate sequence from the gene is amplified and banded on an electrophoresis gel, a female (XX) will generate a single band, and a male (XY) a double band. On the basis of this difference, Gerstenberger established that the skeleton, as suspected, was indeed genetically female. What is more, she was not the only female who had been laid to rest within this apparently male shrine. In all, the amelogenin tests indicated that only six of the skeletons were male, leaving two who were female. If the skeletons beneath the chancel really did correspond to the names recorded on the wall, then one of the earls was missing, and there were two female interlopers in the shrine.

To get closer to their identity, Gerstenberger started work on DNA sequences from their Y chromosome. She focused her attention upon four microsatellite regions along the chromosome. Nearest to the stones commemorating the first two earls, Hanns Cristoph and his son Hanns Sigismund, were two male burials with identical Y patterns, with tandem repeat lengths of 14, 23, 12 and 28 at the four regions studied. Moving to the next generation, two of Hanns Sigismund's sons are commemorated in the chancel. One gave a poor amplification result, but the other, Wolf Ehrenreich, once again came up with 14, 23, 12 and 28. The grave nearest to his son's inscription was apparently robbed, but his grandson Josef Wilhelm followed, echoing the same genetic pattern. So far, so good, but this is where the bones began to deviate from the inscriptions.

We know from contemporary records that Josef Wilhelm was very anxious to have a son and heir, and was rather relieved when Georg Josef turned up. He might, however, have been surprised by what Julia Gerstenberger would later discover. Georg Josef's tandem repeat signature came out as 14, 24, 13 and 33. It is inconceivable that such a switch could have occurred in a single generation, or even in several generations. His true father was not the previous earl.

This left the two female bodies, who obviously could not be studied through the Y chromosome. However, a study of autosomal microsatellites would bring them into the picture. Autosomes (the normal chromosomes) had the advantage of being found in both males and females, but, because they recombined with each generation, could not be used for the kind of lineage tracking possible with either mitochondrial or Y haplotypes. Their careful analysis would allow the women to be related to the male line, and Gerstenberger to trace this story to its conclusion.

The development of the teeth and bones from the two female skeletons indicated different ages at death. One skeleton was of a mature woman of approximately thirty, the other of a girl of around thirteen years of age. What the autosomal microsatellites generated was a pattern consistent with the two females being the illegitimate earl's wife and daughter. In other words, the microsatellites from the skeletons of the male and the older female could be recombined to fit the pattern found in the younger female. We know from historical records that this earl's wife died at the age of thirty-six, matching the skeletal evidence. Both these women had died well before their time. The illegitimate earl had inadvertently broken the male line and now was consciously breaking with tradition. Having lost his wife when she was in her prime, and his daughter when she was still in her youth, he might not be able to give them a memorial inscription but he certainly wanted them laid within his ancestral mausoleum. The next earl chose to be buried elsewhere, and the male line in any case came to an end in 1815. Their ancestors had left behind two stories, the first a formal story inscribed in the chancel walls, the second a real-life story hidden within the DNA. The first story was of a traditional ideal, of a patrilineage of upright, aristocratic men. The second story was of real people, of yearning for an heir, of an extra-marital liaison, and of a funerary tradition defied in the throes of love and grief.

when the names fall silent

How much further could such narratives be taken back in time? In principle, they could go back as far as the recorded genealogies, which in every case is well within the longevity of the short strands of ancient DNA. If it is only the male lines that are documented in full, then more relevant than mitochondrial haplotypes are the Y haplotypes, and research into these is now growing fast. Is there anything to be gained by going back beyond the time scale of these genealogies and examining relationships of kin in prehistory?

The problem might be compared with tree-ring dating and its calibration. The first tree-ring patterns to be fixed to a particular year were comfortably within the historical time scale. They were sequenced records from timber-framed buildings and wooden waterfronts whose dates of construction were securely documented, providing anchor points in the historical record for the patterns recorded in ancient wood. Back in prehistory, these historical anchor points are lacking. At first, prehistoric tree-ring patterns were assembled into floating chronologies, isolated local patterns without an absolute date. In time, these floating chronologies grew to overlap, first with each other, and then with the historically fixed chronologies, and the whole master chronology came together to provide prehistory's most precise dating framework by far.

This is only a loose analogy, and we will probably never achieve comparable master genealogies for prehistory. However, the floating chronology does provide a model of what might be achievable with current methods, a series of floating genealogies, with no actual names to anchor them, but with their interrelationships established. Such an approach might tell us a great deal about past societies through their burial sites. Who was placed in the ground near to whom? Were nuclear families, or larger groups, buried together, or were age and gender more relevant? In Göttingen, Susanne Hummel and Bernd Herrmann decided to look in the nameless archaeological record for something equivalent to the Romanov group. They found it in a cluster of bodies from a burial pit dug in the first millennium AD on the northern Rhine at Kleve, close to the German border with Holland. Excavation had exposed the skeletons of a mature man and woman, neatly laid alongside each other. Between and around them were the remains of the skeletons of three children. The fusion of the children's bones suggested ages ranging between one and ten years. They looked the ideal test for a family burial, to be assayed by microsatellite studies of ancient DNA. Using microsatellites similar to those examined in the forensic cases discussed above, the team established that at least three of the five bodies were close relatives. In a similar approach to that taken in the case of the earls of Königsfeld, they used the ancient DNA to establish both kin relations and the sex of the burials. Putting these together, the five bodies could be interpreted as a mother and father and their three children – a daughter of eight to ten years of age, a son of one or two years, and a third child, whose remains were the only ones too poorly preserved for analysis.

Family structures are always interesting, but especially so when society is in the process of considerable upheaval. At such times the balance between the clan, the extended family, the nuclear family and so on may change as a major and integral part of the wider changes in society. Take, for example, the transition between foraging and farming that has been a recurrent theme in biomolecular archaeology. A mobile hunting group and settled village may organize their families and relations of kin in quite different ways, and understanding those changes is central to understanding the transition as a whole.

Half a century ago, the anthropologist George Peter Murdock looked for a global logic in the relationship between kin patterns and the broader organization of society. Examining a large number of living and recent societies from around the world, he explored kinship relationships, the ways in which property was passed down the family line, and the norms that determined which relatives lived near to which. He went on to relate these patterns to other social and economic attributes of each community. One pattern that captured his attention was a trend for patterns of inheritance and residence to be linked. If goods passed down the male line, then more often than not the male line stayed together in the same village, the wives joining from elsewhere. Conversely, if goods passed down the female line, then it was often the husbands who joined their wives' family villages. In essence, Murdock saw this as a way in which a community would maintain its accumulated stock. It was by no means a hard and fast rule. There were many exceptions and some completely different patterns of residence and inheritance. Nevertheless, there did appear to be a widespread trend, linking patrilineage and patrilocality on the one hand, and matrilineage and matrilocality on the other.

Other anthropologists had seen the same trend as linked to different ways of life. Murdock quotes the words of an earlier anthropologist, R. C. Thurnwald:

> [S]ons inherit the trapping and hunting gear of their fathers; daughters the cooking utensils and food gathering implements of their mothers. When the women have advanced from collecting to agriculture, their property is augmented, and matrilineal inheritance consequently becomes the more important. (Thurnwald, 1932: 193–4, cited in Murdock, 1949: 205)

It was this kind of observation that led to the generalized idea of a shift from patrilineal inheritance and patrilocal residence among hunter-gatherers, to matrilineal, matrilocal 'lower agricultural peoples', changing back again in 'advanced' hierarchical societies. This generalized idea is heavily imbued with rather dated notions, such as a fixed notion of gender roles, and the unilinear evolution implicit in such words as 'lower', 'advanced' and so on. There was little archaeology could do with this idea beyond pasting kinship patterns observed today rather uncritically upon the mute remains from the past. There is still very little kinship evidence from ancient DNA to enable us to explore the issue, but enough to see how the work might unfold. Three of the most detailed studies to date are distributed rather arbitrarily around the world, but by chance they come from a hunter-gatherer community, an early farming society, and a relatively recent pre-Columbian community. Let us consider them in reverse order.

Anne Stone and Mark Stoneking had gleaned several pieces of valuable genetic information from their pre-Columbian burial mound on the Illinois River at Norris Farm. Much of this information has contributed to the story of America's first colonization, discussed in Chapter 7. This burial mound had been associated with a rather short-lived village of the Oneota culture, on a distant periphery of that culture's central range. This is just the kind of context in which we might expect quite a small effective breeding population and a discernible pattern of kin relations. What the researchers actually encountered was a high degree of genetic diversity, and a marked absence of maternal or sibling patterns on the ground. This was, however, based on mitochondrial evidence alone.

They amplified DNA sequence information from over 100 skeletons, and for more than fifty this included detailed information about the mitochondrial control region. For the control region section examined, three-quarters of the individuals carried rare or unique lineages, 50 per cent more than the average rate in modern populations. No sign of a genetic narrowing there. Might men and women show different patterns? So far as Stone and Stoneking could see, they did not. The diversity was high in both sexes. What about the positions of the burials on the mound – might there be family clusters at different points? The major lineages displayed no such grouping. Each of the lineages was found across the mound. The cultural evidence may have been of a community that was remotely situated and perhaps isolated, but the mitochondrial picture was of considerable mixing, a true melting pot of maternal linages.

This is the kind of pattern we might expect from patrilocality, or indeed one of the less common residence patterns, but certainly not from matrilocality. It is not a surprising result; many living Native American communities come from a patrilocal tradition. Where we might expect to encounter matrilocality is among the early agricultural communities of the Neolithic period. These have often been regarded as forming an egalitarian, matrilineal interlude between the early hunters and the later hierarchical warrior societies in which the male line came once more to the fore. The early farmers of Europe left many a collective burial across the landscape, often within a substantial monument of earth or stone, yielding ideal populations for kinship analysis. One such monument was excavated in Calvados in northern France, and scientists at the Institut Pasteur at Lille set about tackling the issue of the relationships between the buried individuals within.

The burial monument constructed for the early prehistoric farmers at Condé-sur-Ifs was made up of six passage tombs, each with around a dozen bodies within, laid to rest 7,000 years ago. In one of these chambers, excavation revealed eleven skeletons laid out in such a way as to suggest two family groupings. Each grouping was made up of a cluster of bodies which included both adults and children. Thomas Delefosse and Catherine Hänni wondered whether they might be able to emulate the study of the Romanov remains, but on a much older set of bones. From the two putative family groupings they took samples from three adults and two children and set about amplifying and sequencing parts of the first and second hypervariable segments of the mitochondrial control region. The three potential outcomes they had in mind were: that both groups were closely related; that the groups corresponded to two unrelated families; or that the individuals in the chamber were not maternally related at all.

What they found was that in the five individuals examined from the eleven bodies in the chamber, in relation to both hypervariable segments all their mitochondrial sequences were distinct. The individuals were clearly not maternally related. This is not to say they were not related at all. However, the mother–child relation that might have been inferred from the arrangement of burials was certainly an illusion. Moreover, they were not sibling groups. Despite the rather cosy grouping suggested by the six groups of around a dozen bodies each, themselves arranged in small clusters, the genetic diversity of those laid to rest within the tomb was clear. The authors admit that this is only the start of the work, but a start which already disposes of some assumptions. What is absent from the Condé-sur-Ifs data is a reflection of matrilocality within the early farming community buried here. As with the much later Norris Farm community on the far side of the Atlantic, the mitochondrial diversity suggests something different. It will come as no surprise that the third assemblage, older still, also displayed significant mitochondrial diversity.

Like the mound at Norris Farm, the ancient sinkhole at Windover in Florida has also figured several times in the short history of biomolecular archaeology. The bodies preserved within it, most of all their extraordinary pickled brains, are evidence of how abruptly decay can be arrested when certain conditions of water, air and acidity coincide. One such brain from the Windover pond yielded one of archaeology's first findings of ancient human DNA. As the number of bodies studied increased, so the pond took on a new interest. It became the site where some of the rarer New World mitochondrial lineages, now almost confined to the heart of the Amazon, were widespread in prehistory. By the mid-1990s, over 170 individuals had been lifted by the archaeologists, and 50 per cent retained soft tissue intact. Bill Hauswirth and his colleagues could now probe yet deeper into the communities that deposited their dead in the Windover pond 7,000–8,000 years ago. As with the other burial studies discussed above, the mitochondrial control region provided an obvious sequence for study, and Hauswirth and his colleagues amplified a 168-base-pair sequence within the control region. Among fourteen individuals sequenced, they found eleven distinct haplotypes, another very high proportion. The Windover population also yielded ancient DNA data so far unrecorded from the other two burial groups discussed above.

The reader may by now be wondering why several of these kin studies have looked at the mitochondrial patterns alone. There would seem to be an obvious pairing between mitochondrial studies tracking female lines, and Y chromosome studies tracking male lines, and this will indeed form the basis of ancient kinship studies in the future. What we see in past work is the parallel unfolding of modern human genetics as a science. Information about the mitochondrial genome came on stream ahead of other parts of the genome, and our detailed knowledge of the Y chromosome has only recently expanded to its current state. Anne Stone is keen to get back to the Norris Farm individuals to examine Y sequences, but that is in the future and subject to the consent of the relevant living communities. The same is true of Windover Bog, but what that population has yielded is some autosomal information.

In addition to examining mitochondrial sequences, Bill Hauswirth looked at other parts of the genome, less frequently targeted in ancient DNA analysis but dovetailing with a major strand of modern human genetic systems. One of the most valuable genetic markers among the blood proteins is a system of proteins on the surface of white blood cells, part of the body's armoury against invading disease organisms. These proteins, the human lymphocyte antigens, or HLA for short, are probably the most variable protein system within the body. Indeed, a variety of defence systems is central to ensuring that at least some people survive to reproduce however heavily disease hits. That great diversity has incidentally been of enormous value as a marker of human genetic diversity, and has been central to Cavalli-Sforza's genetic synthesis. The DNA blueprint for HLA is to be found on an arm of the sixth chromosome in the nucleus as a string of genes, each one with ten to fifty possible alleles. Hauswirth decided to extend his study from the well-worn path of the mitochondrial control region to the nuclear genome, and to chromosome six. He picked a hypervariable, non-coding region within the HLA encoding arm of the chromosome, and charted its diversity within a sample of fourteen individuals. The diversity was high, another reflection of the mixed gene pool that first entered the New World, but one particular element caught his attention.

At one specific locus, all except one of the individuals shared one allele in common. In many a less variable region this would not merit comment, but in this highly variable part of the genome, it did suggest a blood connection between the individuals. This possibility Hauswirth was able to explore by looking at some other highly variable regions. Like many interested in recovering ancient kin patterns, he turned to microsatellites. Hauswirth and his colleagues focused upon one particular gene on the first human chromosome. This gene contained one of those microsatellites that Hagelberg had exploited, made up of tandem repeats of the cytosine–adenine couplet. There may be fewer than ten, or more than twenty couplets in the microsatellite, according to genetic lineage. Among the Windover individuals, however, two particular microsatellite lengths dominate the population, one of nine repeats and the other of fifteen repeats. Alongside the HLA evidence this was further indication of close blood ties within a population placing its dead in the pond over several centuries.

If we step back from HLA genes and microsatellites to think of how this waterhole was used by living people, it is worth remembering that a sinkhole is not an uncommon feature in the Florida landscape. A mobile population of hunter-gatherers roaming a sparsely settled continent 8,000 years ago could and would have stopped by several of these. Yet this one had a special significance for a particular community. It was not just where they rested, drank, and procured their food. It was also where they offered up their dead – not only more than 8,000 years ago, for their direct descendants were still being offered up to their watery grave 1,100 years later. It is as if we were to walk into our local graveyard to find the surnames of our family and neighbours, but on headstones that had been gathering lichen since the days of Alfred the Great and Charlemagne.

Of equal interest is the fact that this continuity fails to show up in the mitochondrial data. Hauswirth and his colleagues amplified a 168-base-pair sequence within the control region. Among fourteen individuals sequenced, they found eleven distinct haplotypes, another very high proportion. The most distinctive evidence of relatedness came from autosomes, inherited by both sexes, while the mitochondrial evidence, inherited through the maternal line alone, displayed a now familiar level of diversity. The visitors to the Windover Bog displayed the genetic hallmark of patrilocality.

Once combined studies of mitochondrial, Y and autosomal lines become more frequent, we will no doubt encounter a number of different patterns. We have seen in an earlier chapter how the contrast between mitochondrial and nuclear evidence from both Europe and the Pacific Islands might imply that, in these particular cases, the female lines were less mobile than the male lines. What is significant is that ancient DNA has the potential to discriminate, a potential that will continue to sharpen, both in relation to the particular kin groups and to those who join them from the outside world.

strangers in their midst

The modern world is full of outsiders. Today's cities are geared up to accommodate the stranger, whose money can be exchanged for food, shelter and some kind of instant social life. In the great majority of prehistoric societies, true strangers would have been different indeed. Their entry into a remote village would have been subject to the sanction of the senior members, and under the curious gaze of everyone. These were communities interwoven by kinship and familiarity. Without money, the transactions they made would have been embedded in an intricate fabric of social relations. Ancient DNA may be able to bring elements of that fabric to light, and to illuminate the occasional stranger within it.

There is an ancient mound of discarded shells, at a place called Takuta-Nishibun on the Japanese island of Kyushu, that was home to many generations of prehistoric fishermen and foragers. Just over 2,000 years ago, this ancient place of food preparation and shelter was transformed into a place to bury the dead. These were the dead of a community that followed what was then a novel way of life. This new generation farmed rice, and their dwelling area had shifted from the ancient mound to a small adjacent village. From their burials in the ancient shell mound, we can tell something about this small community of rice farmers, and their links with a wider world.

That world stretched out in two directions. In one direction was the Asian mainland. It was from that direction that rice farming had arrived in the not too distant past. Travellers continued to make the crossing, as we can tell from the imported mirrors and objects of metal and glass that accompany the burials on the shell mound. Stretching out in the opposite direction was the Japanese island string, where a different tradition could be found, the much older tradition of fishing and foraging that stretched back 10,000 years. Many have argued that these two traditions were quite distinct, reflecting two different 'peoples' with different ways of life, different ancestries, and even different skull shapes and skeletal forms. On this site they seem to be intermingled, an impression backed up by studies of their DNA.

In a genetics laboratory at Shizuoka, not far from Tokyo, geneticist Horishi Oota extracted DNA from twenty-five skeletons from the Takuta-Nishibun site, together with two other skeletons he had acquired from a nearby site of similar date. Previous researchers had managed to construct from DNA evidence an evolutionary bush, drawn from the entire Japanese island string and neighbouring parts of east Asia. It provided a framework within which Japanese farming populations might be distinguished and separated from pre-agricultural foraging populations. The twenty-five Takuta-Nishibun individuals failed to comply with such a neat categorization. The scatter of their haplotypes across the entire evolutionary bush reflected, not a neat replacement of foragers by farmers, but instead a tangle of intertwined histories, reflected by the close association between farming village and shell mound. That is not to say the village dwellers saw no divisions; they clearly did.

There were two distinct burial traditions on the mound. Most bodies were placed directly into the mound, but about one third of them were laid to rest within large earthenware jars, one of the commoner practices among early rice farmers in these parts. Oota examined nine individuals buried within jars, and seventeen who had been interred directly within the shell mound. From them, he amplified DNA from the third hypervariable segment of the mitochondrial control region. The bodies interred directly within the mound turned out to be genetically diverse. The seventeen individuals grouped into ten mitochondrial haplotypes. Their internal variation was such that Oota could argue for a distant common ancestry. They were not the descendants of a single lineage of recent arrivals. When he turned his attention to the jar burials, the pattern he found was quite distinct. All nine sequences were within two base-pairs of each other, and six of the nine sequences were identical. Here was a much narrower genetic grouping of individuals, related through the female line and displaying their distinct family identity through the way in which they buried their dead, and no doubt through other rites of passage that evade the archaeological record. This division was not hard and fast. Three of the haplotypes were shared by both burial traditions, and this includes the commonest haplotype among the jar burials, accounting for two-thirds of their number. Against a background of two distinct traditions, there was mixing and intermarriage. That mixing occasionally incorporated individuals from further afield.

The excavators came across the body of a young girl placed, not on the mound, but rather closer to the rice farmers' village. When Oota came to examine her mitochondrial sequence, he discovered she was set apart in another way. Her sequence was distinct from that of all the other villagers, and placed her on a rather distant arm of the Japanese family tree. Who was this young girl, and where were her relatives? The Takuta-Nishibun site offered no answer, but there is another archaeological site that does. Just six kilometres to the north of the shell mound, the bodies of a woman and a girl had been laid to rest in earthenware jars, half-way up a hill called Hanaura. Each body was amply adorned with shell bracelets. Perhaps this, and their lofty position on the hill, reflected an equally lofty status in life. The excavators surmised that they might be mother and daughter from a family of shamans. Their DNA indicated that things were a little more complicated.

The woman and girl were not mother and daughter; their mitochondrial sequences were very different. Furthermore, the woman had precisely the same sequence as six of the nine individuals buried in jars six km to the south. She was essentially part of the same lineage and shared in its burial tradition. The lineage of the girl who was similarly buried and similarly adorned was unusual, but not unknown. Her sequence precisely matched that of the stranger in the south, the young girl buried separately from the other villagers.

That ancient community on Kyushu Island was already an amalgam of two communities living together. One traced its ancestry back to an ancient time when the sea and its bounty were the basis of life, and the vast mounds of discarded shells upon which they dwelt and buried their dead a symbol of that ancient tradition. The other community traced its ancestry back to a more recent history of migrating rice farmers, coming from a land they still visited for trade. But the latter community intermarried with the first, lived in the same settlement and shared its burial mound. Their physical and genetic differences became blurred and confused. At some stage, two girls arrived from distant parts, perhaps to be betrothed. Both died young, but not before one was fully assimilated, honoured with the prestigious burial accorded to her new family. The other died before any such assimilation. At some distance from the main cemetery, she was buried alone.

This ancient village in Kyushu demonstrates the two features of prehistoric demography that in so many of the examples explored here are found to co-exist strangely. On the one hand, certain traditions, rooted in the paternal or maternal lineage, can persist for countless generations, creating genetic and cultural patterns that stay more or less intact for hundreds or even thousands of years. Nevertheless there are, within this framework of considerable persistence, countless opportunities for moving beyond the immediate circle of friends and relations, leading to some significant elements of mixing on a far larger geographical scale. Sometimes the exchange of artefacts over long distances allows us to chart the possible routes along which this mixing might have taken place. At other times, as at the Takuta-Nishibun site, biomolecular evidence can actually introduce us to those newcomers who have travelled from afar. However, DNA is not the only biomolecule that can track these individual travellers.

isotopic traces of the traveller

The Chinchorro mummies of the South American coastline provide a rich source of ancient biomolecules beyond DNA alone. The extreme desiccation of the coastal strip has conserved many of the bodies' proteins, such as the collagen of their bones and the keratin of their hair. The isotopic signals in these confirm just how dependent on the sea the Chinchorro societies were. From their coastal villages they looked back to the rocky slopes rising up to the distant Andes, and saw one of the world's most parched and barren land surfaces. Facing in the other direction they saw before them a fertile and productive ecosystem, the shallow coastal waters beneath the pleasant warmth of a Pacific sun. No wonder their middens were stacked up with the bones and shells of fish, molluscs, sea-lions and coastal birds, with barely a trace of terrestrial foods. Consequently, the isotopic signal in their proteins is substantially marine, a feature that shows up markedly in the isotopic ratios of carbon, nitrogen and sulphur. Steve Macko established this much from hairs taken from the mummified bodies from the early prehistoric coastal site of Morro, one of the sites discovered when Max Uhle first brought the Chinchorro mummies to light. The isotope ratios for carbon, nitrogen and sulphur all clustered tightly in the range expected from a marine diet.

By way of contrast, Macko went inland, following one of the fertile green valleys that dissect the barren hills on either side, to the remains of an ancient inland settlement. The margins of these valleys still allow the mummification of corpses and so he was able to collect hair from these also. Here, the isotopes generated a contrasting picture, indicating a diet dominated by terrestrial rather than marine foodstuffs. Although not far apart, the two settlements were separated into distinct food chains, one exploiting the coastal waters and the other the valleys. Their hair retained an isotopic imprint of this lifestyle, and that is how Macko managed to track down movements between them.

The hair of just one of the inland individuals carried an unmistakably marine signal for each of the isotopes. He died inland, but coursing through his bloodstream was the mark of his coastal diet. He had clearly arrived recently from the coast. As we have seen, the hair responds quickly to a changing diet. Yet he was not sent back home to be buried, but carefully laid to rest at his destination – a direct record of contact and movement between two discrete communities.

making contact

All this is still a far cry from placing a name and identity upon a set of bones, as in the case of Joseph Mengele, the 'Angel of Death' of Auschwitz. Nevertheless, it is not so far from the recognition of the Romanov family, the study upon which a number of attempts at prehistoric kinship analysis were modelled. It is already clear that the archaeological record will not generate a faithful and unwavering mirror to reflect some later scholar's cross-cultural patterns, though trends may be there. Such trends as exist, reflecting the changing norms of ancient communities, will be overlaid with evidence for subversion of and deviation from those rules. In addition there will be interlopers to the burial group, travellers from afar. The study of ancient kin is in its infancy, and will no doubt throw up many surprises, especially when mitochondrial and Y haplotypes are examined together. It will also begin to describe the complex interweaving of social networks on larger and larger scales, which may well prove to be the principal story of human prehistory. For a long time those growing networks were hinted at by the sourcing of precious artefacts that had travelled vast distances from their place of origin. Now, with ancient DNA, we can come into contact with families, communities, and the actual individuals who travelled between them.

10

enemies within

ways of the flesh

Maria of Aragon was a woman of substance in Renaissance Italy. As Marchioness of Vasto and a personal friend of Michelangelo, her social status during life earned her in death a place in the Abbey of San Domenico Maggiore in Naples. After four centuries in the abbey, her body was disinterred. Gino Fornacieri, a palaeopathologist at Pisa University, had gained permission to analyse her mummified remains. Having noticed signs of disease on the body, he in due course dispatched samples to Franco Rollo at Camarino University. Rollo had been applying the new DNA techniques to a variety of unusual ancient tissues, and so was the obvious person to check out Fornacieri's diagnosis. The samples included part of a linen patch, filled with ivy leaves, that had been found on the Marchioness's left arm. On the skin beneath the patch an oval ulcer was still visible. With his scalpel Rollo carefully eased off some cellular tissue, and began his search for traces of ancient DNA. What he found allowed him to confirm Fornacieri's diagnosis of *Treponema pallidum*, the agent of venereal syphilis. Yet another black box in the archaeological record had been prised open by molecular analysis. Ancient DNA would open the way to study a range of hitherto invisible diseases and their pathogens.

The woman's body leaves much to the archaeological record, while the syphilis bacterium is as difficult for us to see as it was for her. Yet both organisms contain a genetic blueprint that can be amplified to a similar volume. As well as enriching our understanding of organisms from the archaeological record that we can see with the naked eye, the molecular signal can be as strong for something microscopic as it can be for a large mammal. As Rollo's work was demonstrating, molecular techniques could bring into view the whole range of organisms involved in past human life, whatever their scale.

In working with the bacterial genome of syphilis, Rollo was dealing with a looped DNA molecule not dissimilar from those within the mitochondrion and chloroplast. Indeed, it is now believed that those sub-cellular structures first arrived in ancient plant and animal cells by the invasion of something rather like a bacterium. Like the genomes of those cellular bodies, most of the syphilis genome is contained within a circular DNA molecule, but a rather larger circle. It is about ten times the size of the chloroplast genome, and about 100 times as long as the human mitochondrial genome, and it contains a few thousand genes. Other bacterial genomes have varied forms. Some are doubly and triply coiled about themselves in order to pack into tiny cells. They also have variable regions in common, which can be used for identification and phylogenetic analysis.

The recognition of these bacterial genomes has largely relied upon a gene involved in building ribosomal RNA, known as the 16S rRNA gene. Something rather similar is found in the mammalian mitochondrion, where it serves as a useful molecular clock. All organisms need such a gene, as part of their RNA equipment, and the 16S is simply a way of describing the physical size of the gene. Within bacteria, this gene varies in sequence structure and serves as a useful identifier and evolutionary clock for bacteria. Using this gene to identify what else was in the dressing, Rollo encountered a range of other bacteria normally found living inside our mouths, part of the natural wildlife of our saliva. The Marchioness had had her ailment treated by what contemporary doctors described as 'salivation cures' for the *morbus gallicus*.

Retrieving such a detailed medical history for a named public figure from the past has its own slightly troubling allure, but getting to grips with the range of micro-organisms involved in the human past has a much larger, less anecdotal significance. We know surprisingly little about patterns of human health in the very long term.

We can say a lot about changes in health and medicine during the last 200 years, and to some extent as far back in time as the ancient Greeks. The earliest 99 per cent of our species' history is, however, far more open to conjecture. It is clear that in recent times scientific medicine has eradicated or minimized certain diseases, and reduced mortality at certain key points in life, in particular during childbirth and the first few years of childhood. We can also contrast a period of improving health in the modern period with the horrors of the Black Death in the Middle Ages. There is growing evidence of something in the sixth century AD rather similar to the Black Death, and known as the Justinian Plague. Beyond that, we are heavily dependent on what we can glean from ancient skeletons, and they are open to contrasting interpretations.

One view would have the improvements of recent history as the summation of an extended period of progress, in step with the overall progress of human history. This view sees the progress in knowledge and invention as increasingly buffering us from the worst that nature can offer. A converse view would see the historical records of pre-twentieth-century health as the end point of a long period of decline. That decline is seen as following our Neolithic deviation from nature, and our propensity to crowd large numbers of stressed and under-nourished people and livestock into compact permanent settlements. The more extreme that deviation, the more of a breeding ground we ourselves became for opportunistic micro-organisms.

Conventional archaeological evidence does not easily discriminate between these two views of the past. Prior to the twentieth century, human populations of a variety of periods and places yield evidence of the high levels of infant mortality from which modern medicine has released some modern populations. However, many of those who survived beyond childhood stood a good chance of reaching three score years and ten with a good set of teeth. The low life expectancies computed from ancient population evidence are a consequence of including infant mortality in the global average. The skeletal evidence itself leaves many questions about long-term trends in health unanswered.

This uncertainty about our past has implications that go beyond the actual experience of sickness and health. Micro-organisms are our principal predators, the next step up in our food chain. The dynamics of our population patterns through time are as much bound up with the micro-organisms that eat us as they are with the species available for us to consume. Nothing illustrates this better than the cyclical repetition of epidemics, or the pandemic that came to be known as the Black Death. Between the years 1347 and 1351, this voracious pandemic reduced the population of Europe by a third, and precipitated a population decline sustained long after the disease itself had subsided. The much more fragmentary evidence from the earlier Justinian Plague suggests that this too could have been a sustained and pernicious killer.

These two pandemic diseases clearly had a profound impact on life, death, and the sheer scale of the human population of Europe over the last 2,000 years. As a consequence, the pattern of population growth for the two millennia preceding the Industrial Revolution looked very different from what we have seen since. Early historical Europe underwent, not the unstoppable rise of industrial populations, but long, slow cycles of growth, disease and decline. Looking further back beyond the written record, we have two models from which to choose. One model treats what we see today as the norm, albeit on a different scale. Human population has an innate tendency to grow exponentially, particularly since the beginnings of agriculture. The disease episodes are particular events lodged in particular histories of urban squalor. This is the model that implicitly underpins the concept, discussed in Chapter 7, of a number of expansions and journeys. The alternative model treats modern growth as an exception, an artefact of industrial societies. The prolonged mediaeval oscillation of a human population, caught between its prey and its micro-predators, provides a better fit for settled farming communities in prehistory.

Without knowing what the dynamics of our micro-predators are through time, we have no reason to assume they are more like the exponential rise of recent centuries than the drawn-out oscillations discerned from late prehistory onwards. We need to know whether the Justinian Plague and Black Death are unique to the period of documented history. Are they features of dense, urbanized societies only, or is the long saga of human prehistory peppered with surges of micro-predation, repeatedly curbing populations? The archaeological record is fairly silent on the matter. Even though we can look at ancient skeletons, examine their teeth and bones, assess their life-spans, and sometimes encounter bones that are clearly diseased, the archaeological record gives us no easy answers. It does not help us discriminate between these two stories. We can see why, when we take a closer look at one of the more meticulously studied cemetery assemblages from the time of the Black Death.

life and death in mediaeval york

On the outskirts of the City of York in northern England, the ruined church of Helen-on-the-Walls was fully excavated in advance of a housing development. Also unearthed were the bodies of well over 1,000 of the parishioners who attended the church between the tenth and sixteenth centuries. We know a great deal about that mediaeval population of this parish of York, thanks to the scientific analyses undertaken by the archaeologists. They revealed that life was hard, with one in four of the parishioners dying in childhood, and with women of childbearing age also vulnerable. Only one in ten of either sex reached the age of sixty. We can infer how long men and women could expect to live, how well they were fed, the extent to which their working lives made their bones ache, and so on. We can also say something about their health. One had bone lesions resulting from syphilis; tuberculosis had collapsed the spine of another. Many of the remaining skeletons displayed symptoms of arthritis, but in general their health did not appear to be at all bad.

That is quite an interesting comment bearing in mind the dates of the burials. They span that period of the later fourteenth century when a third of all people in Europe died as a result of the cyclical resurgence of a rat-borne bacterium called *Yersinia pestis*. York suffered the Black Death with the rest of Europe. The parishioners of St Helen's were presumably not exempt, and yet an intensive study of the bones of 1,000 of those parishioners revealed not a trace of the killer disease. The reason is simple. The faster a disease kills, the less trace it leaves on the bones. For a disease to show up on our skeletons, we have to stay alive long enough for the bones to suffer visibly. For the spine of a tuberculosis victim to collapse, for the skull of a syphilitic to become riddled with lesions, for the facial bones of a leper to erode away, life must go on. The main disease we see in archaeological skeletons is arthritis, the wear and tear on the joints, which may give considerable discomfort and pain, but which generally allows the sufferer to survive for many years. In terms of archaeological evidence, fast killers like the Black Death fall into a large black hole.

If these skeletons bear no mark of the greatest trauma of contemporary mediaeval life, how many other prolonged plagues and pandemics have remained invisible before written records? The Black Death is an unusual episode, but we know from accounts of the Justinian Plague that it is not unique. As with the Black Death, contemporary records document the devastation of the Justinian Plague. However, the innumerable bones from the many late Roman graveyards retain no discernible trace of the disease.

The first steps to greater knowledge have now been taken by molecular archaeology, not to seek out symptoms on a blemish-free skeleton, but to find the disease itself. Just as Franco Rollo had sought out the DNA fingerprint of the syphilis bacterium in a sixteenth-century Marchioness, Michel Drancourt and his colleagues at Marseilles embarked upon a similar search. They were looking for the plague bacterium at the heart of the Black Death, *Yersinia pestis*.

A starting point for the studies of Drancourt's team was to find some skeletons with a clear association with the plague bacterium. These they found at two mass graves in Provence, France, which documentary evidence had linked to plague quarantine hospitals. At Lambesc, 133 corpses had been placed in such a pit in 1590, and at Marseilles, 200 bodies had been buried in a single month in 1722. They selected the jaws of six of these and, in order to minimize the chance of DNA contamination, worked with the internal dental pulp of unerupted teeth, probably the part of the skeleton best protected from the outside environment. Powdery remnants of the pulp were scraped out and subjected to PCR using primers specifically designed for *Yersinia pestis*. For comparison, seven other skeletons were similarly sampled, this time from a mediaeval graveyard in Toulon, with no specific documentary connection with plague. None of these Toulon samples produced a PCR product, whereas three of the six plague-pit specimens did. For the first time, ancient plague had been confronted in the archaeological record.

At the time of writing, this has not been taken further, but work to do so is under way in Oxford. Across Europe, there are many burials coinciding with the Justinian Plague that would be suitable to test. The most interesting goal would be to establish when and where the bacterium prevailed in prehistory. It may have been something relatively new in classical times, a consequence of the sheer scale of trading links in the ancient world, bringing so many together from distant regions into large classical towns. On the other hand, in an agrarian world in which clustered permanent settlements were the norm, these pandemics may always have been a possibility whenever some external factor shifted the sensitive balances between micro-predator and human prey. On the basis of ancient tree-ring evidence, Mike Bailley has argued that the climatic perturbations that might encourage pandemics are scattered through prehistory. DNA may now begin to provide an answer.

the white death

The plague bacterium has come to and gone from the Old World population according to a long-term dynamic, one which we have yet fully to comprehend. Only when the search for *Yersinia pestis* has reached back into prehistory will we be able to do so. Some other diseases by their very nature seem to fit into a long-term trajectory of the human past.

The ancient Egyptians described a disease that was almost certainly tuberculosis, as did Hippocrates, the Greek forefather of medicine. Tuberculosis, sometimes known as the 'white death', is a disease the spread of which would be directly stimulated by the central features of Childe's agricultural and urban revolutions. Farming brought the lives of humans and animals close together, and cattle are a likely source of the disease. Close proximity between townspeople further assists the bacterium responsible for this illness. In historical times, up until the twentieth century, tuberculosis has been a faithful and fearful companion of the progress of industrialization and urbanization around the world. Infirmaries for the sufferers of 'consumption' became as familiar a feature of nineteenth-century cities as their smoke-filled skies.

The disease gives the impression of being an artefact of the Neolithic deviation from nature. Particular strains of the very hardy tuberculosis bacterium are passed easily between humans and cattle, particularly if the blood or the milk of those animals is consumed. As the relationship between humans and domesticated cattle has grown increasingly intimate, so has tuberculosis flourished. This much can be gleaned from recent medical history, and by records showing how, in recent centuries, the spread of tuberculosis accompanied the spread of western agriculture across the world. It has not simply been a marker of the proximity of humans and cattle, but has also flourished where humans themselves live extremely close to each other, especially where overcrowding has been accompanied by malnutrition and poor hygiene. While industrialization and urbanization moved steadily forward in the eighteenth and nineteenth centuries, tuberculosis reached near-epidemic proportions, and was the leading cause of death in the western world. Infirmaries were built to house the sick, but sometimes the crowding and poor living conditions in these places of refuge further encouraged the spread of this poorly understood disease. Late eighteenth-century records of one such infirmary in Newcastle, in the north of England, show that one in ten of the patients admitted suffered from tuberculosis. By the time the patients within that infirmary expired, the prevalence of the disease had almost trebled. The sailors, pitmen, labourers, soldiers and industrial workers who were admitted to the infirmary in great numbers met with the overcrowding and malnutrition upon which the disease relied.

With the early development of agriculture came the shrinkage of the food chain to a small number of domesticated species, all concentrated together with the human population in compact, permanent settlements. It has been seen as the ecological precondition for population growth and complex society. It also seemed to be the ecological precondition for diseases such as tuberculosis, and for the widespread suffering they inflicted. This progress of tuberculosis through the ages, running in parallel with the progress of agriculture itself, can be followed in the archaeological record through the skeletal deformations suffered by some of the afflicted. One particular pattern of bone deformation was very characteristic. An eighteenth-century physician, Sir Percivall Pott, was the first to describe the pattern of eroded spinal discs and collapsed vertebrae that gave its sufferer an angular spine and a hunched back. The condition is still known as Pott's disease. Similar bone deformations, which may have resulted from tuberculosis, have been found scattered through the archaeological record, dispersed along the path of agricultural and urbanized societies.

One of the first Neolithic farmers to open up fields near to what is now Heidelberg in Germany suffered from the disease. By the Bronze Age, deformed bones are accompanied by pictorial representations of hunchbacked sufferers of Pott's disease, and by the beginnings of the historical period, skeletons with the characteristically collapsed spine are found as far to the east as Japan. This spread of evidence mirrors the close relationship between tuberculosis and settled life, farming and the husbandry of domesticated cattle. When Europeans transformed human settlement in the New World, the virulence of tuberculosis became global. If, however, we look back beyond the European discovery of America, we become aware that this is not the complete story. Something else was going on, prior to that major phase of globalization. When Columbus's ships reached America five centuries ago, they may well have brought a range of diseases, but this mass killer seemed already to be on the loose.

Once again our information comes from the fringes of the Atacama desert in Chile, where the particular environmental conditions have led to the remarkable preservation of a group of mummified individuals of the pre-Columbian Chinchorro and Inca societies. One of these individuals was a young girl in her early teens, laid to rest 1,000 years ago in a simple shirt and sandals. One aspect of her garments revealed something significant about her posture. A solid sash and belt had been made to support her lower back, and when her bones were studied, it was clear why. They were as porous as a piece of Swiss cheese. She must have been in great pain, a pain alleviated by chewing coca leaves. She still had a few of these narcotic leaves lodged in her cheek cavity when she died.

This young girl was one of several people found as bodies or skeletons in South America who have the skeletal symptoms of tuberculosis, but she presented something of an enigma. The cattle, and the Old World farmers who we had assumed carried the disease, had yet to arrive in the New World. They would not do so until several centuries later. Her radiocarbon date was AD 1040 ± 70, four to five centuries before Columbus's journey. Other New World bodies take the disease even further back, to as far as 800 BC. If the supposed link between Old World animal husbandry and tuberculosis holds, how could tuberculosis have reached pre-Columbian America in this way? One major limitation of these data is that none of the bone deformities is totally specific to tuberculosis. It could have been argued that the New World deformities had some other cause – that is, until the possibility of detecting the bacterium itself arose.

Of all the disease organisms that might be tackled using ancient DNA analysis, the *Mycobacterium* genus, which includes the species responsible for tuberculosis, has received the greatest attention. A diagnostic stretch of DNA sequence has been identified, a repeat of an insertion occurring several times across the genome and labelled IS6110. By 1993, Mark Spigelman had published a successful amplification of *M.tuberculosis* DNA in a variety of archaeological skeletal remains. They came from a wide variety of contexts, including Byzantine Turkey, mediaeval England and seventeenth- to eighteenth-century Scotland. At around the same time, Bernardo Arriaza's team had begun looking for disease among their rich assemblage of Chinchorro and Andean mummies. They tested for tuberculosis DNA within the vertebrae of the young teenage girl whose suffering from Pott's disease was so clear. She too gave a positive PCR product for the bacterium. Further to the north, in the Osmore Valley in Peru, another group had succeeded in amplifying tuberculosis DNA from the lesion in a mummified woman's lung. The case was strengthening for pre-contact tuberculosis in the New World.

Perhaps the link between the Old World Neolithic revolution and tuberculosis was not quite as clear as it at first seemed. Instead, it is more likely that tuberculosis arrived in the New World with the first colonizers from Beringia. We now know they brought domesticated dogs with them, but certainly not cattle, which had yet to be domesticated and in very different parts of the Old World. It need not have been the humans themselves who brought tuberculosis across, but instead some other mammal, such as buffalo, elk, moose or deer, the populations of which were also expanding into America. Either way, it must have passed between wild animals and their human predators, rather than depending upon the Neolithic deviation and the artificial intimacy of a permanent farming settlement.

When amplifying the DNA, not from the host organism itself but from a separate microbial species that may have infected it, we have to be doubly careful about contamination. Some close relatives of the tuberculosis bacterium are free-living species within the soil. Furthermore, there may well be *Mycobacterium* relatives that are now extinct. The immense detective power of the polymerase chain reaction could be deluding us here – we could be making too much of a tiny number of DNA molecules. Those molecules might even belong to a close relative of tuberculosis rather than to tuberculosis itself, and perhaps we needed some other route to this ancient disease.

from the inner nucleus to the outer coat

The bacterium at the heart of tuberculosis is not simply a package of DNA, but has a fairly complex cell wall designed to defend itself on the perilous journey between and within hosts. The thick wall that guards against desiccation and chemical attack is built of sugars, proteins and long fatty acid molecules packed together. The fatty acids are quite distinctive. Whereas the more familiar fatty acids are built around a single linear carbon chain, these 'mycolic acids' have a complex, branched structure that varies from one type of bacterium to another. They are, in other words, markers of the particular bacteria present. If any of these bacteria leave their DNA behind in the bones, then there is a fair chance they will do the same with their lipid coats. As we saw in the previous chapter, lipid residues can be extremely persistent in archaeological materials, thanks in large part to the inability of the soil water to dissolve them and wash them away.

At Newcastle, Angela Gernaey embarked on a search for these characteristically branched mycolic acids in ancient bone. She had the good fortune to be around when the burial ground attached to the Newcastle Infirmary described above was being excavated, in advance of a new building development. The excavation was a telling experience. So many of the sailors, pit-men and industrial workers who spent their final days here were clearly disposed of, when dead, hurriedly and without much ceremony. They were stacked up, without rows or plots, in a crowded grave space, sometimes in a coffin, sometimes not. Certain of these skeletons had the physical appearance, the collapsed spine or bone lesions, that we associate with tuberculosis. Those were, however, just the tip of a much larger iceberg of infection, as Gernaey was to discover.

She selected a rib bone from each individual from a sample of twenty-one. One had suffered from a collapsed spine, but the others showed no skeletal sign of infection. The sampled rib bones were ground up and their lipids dissolved and slightly modified in order that they could be assayed using a technique called High Performance Liquid Chromatography (HPLC) that sorts molecules according to their mobility through a carefully designed liquid medium. Gernaey built an HPLC fingerprint for each of the twenty-one ribs, and then a separate 'control' fingerprint from a clinical sample of modern tuberculosis, and an extensive series of samples of soil was taken from immediately adjacent to the excavated ribs. The different fingerprints were compared, and the result was clear. The soil samples completely lacked any of the lipid peaks displayed in the control sample. The same was true of sixteen of the rib bones. For the other five, Gernaey found a clear match between the ancient samples and the modern. Just as around one in four patients is recorded as having died of tuberculosis, the rib bones of one in four of the buried skeletons have retained part of the disease bacterium's lipid coat. The lipid has survived in great enough quantities to be detected centuries later.

DNA, lipids and the different types of physical bone lesions are each telling us something slightly different about the disease within these ancient skeletons. The DNA indicates the presence of the disease organism itself, in however small a quantity – the potential of infection. Ancient DNA alone cannot itself demonstrate the onset of disease. The amplified sequences from a small number of bacterial cells will appear similar to the amplified sequence from a heavy infestation. The lipids, however, are not amplified and, judging from the Newcastle Infirmary, are detectable when the bacterium is rife within the bone. Neither the presence of the bacterium nor even intense infection need lead to bone disorders. Indeed, the impaired breathing from damaged lungs can end life before any change to the bones has occurred. Lesions within them indicate the disease within a living person, coping with increasing disablement but staying alive for a while at least. In the St Helen's cemetery in York, only one of almost 1,200 bodies studied had visible signs of tuberculosis in his or her bones. If we were able to return to the bodies armed with a range of molecular analyses we could no doubt place that individual in the wider context of a larger population experiencing the disease at its many different stages.

Mark Spigelman and Angela Gernaey brought their two approaches together via a trace of mineral deposited within an otherwise healthy looking skeleton. Be'er Sheva University had been excavating a Byzantine church in the Negev Desert in Israel when they encountered a 1,400 year old burial. As they examined the skeleton, specifically its ribs, they noticed a limy residue, corresponding to the place the pleural cavity had been. Mark Spigelman recognized this as a fragment of calcified lung pleura, which would suggest that the buried individual had been infected with tuberculosis. No one had ever examined such material for biomolecular evidence, and Spigelman was taken by the idea of bringing together DNA and lipid approaches to test his hypothesis. The calcified fragment came up positive on both counts. It had retained both the DNA and the mycolipids specifically associated with *Mycobacterium tuberculosis*.

enemies or friends?

Micro-organisms have played a major role in human prehistory as our most significant predators. As agriculture has substantially raised the human population so, from time to time, have our microbial predators felled it. Yet not all microbes are as bad for us as the title of this chapter might suggest. The surfaces of our bodies, and in particular our mouths and guts, are full of microscopic wildlife without which our health would become very poor. They aid in digestion, protection against unwelcome microbes, and the provision of some nutrients. The flora of our gut is as much a cultural artefact as that of our back garden, and largely a consequence of what has been fed into it. Different culinary traditions in different parts of the world have tended to generate regional differences in the human gut flora. Having caught a glimpse of the inhabitants of the saliva applied in some manner to the Marchioness of Vasca's left arm, Franco Rollo realized that he could probe yet further. He could examine these ancient floras, and get some sense of how they had evolved through time.

He started probing into another mummified body, that of a young woman from the ancient Inca capital at Cuzco in Peru. The body had withstood the centuries with many internal organs intact. The stomach, lungs, intestines, liver and heart were all recognizable. Rollo's team took samples from the muscle of the heart and from the large intestine, and prepared them for amplification of a short section of the 16S rRNA gene. After various modifications of their method, they came up with a protocol that worked, and generated a series of DNA amplification products that could be cloned, sequenced, and matched to a wide variety of contemporary gut bacteria.

The commonest types of bacteria encountered were of the genus *Clostridium*. This is a widespread genus of rod-shaped bacteria, found in soil and water as well as in the gut. Some species lead to diseases such as botulism and tetanus, but the species within the Cuzco woman were largely benign members of her gut flora. This dominance of *Clostridium* was probably not, however, a fair reflection of the wide range of microbes that would have originally colonized her gut, but one sign of a level of persistence in certain microbes that is truly remarkable.

life in old fossils?

Clostridium species today display a considerable resistance to heat, desiccation, toxic chemicals and detergents. This is because they are among the bacteria able to form spores, one of the most resilient resting stages known in the living world. At the heart of these spores is a condensed package of the bacteria's DNA, bound together with short protein molecules. This condensed package is enclosed by several layers of protective coating, which shield their precious cargo from heat, light and chemical damage. Not much is known about how long these tiny capsules of genetic information can last, but the bias in the Cuzco woman's intestine towards this genus is very likely a consequence of the survival powers of these spores. That power to survive could, however, take us very much further back than the ancient Inca empire, perhaps to the oldest fossils probed in the quest for ancient DNA.

The first three years of the 1990s was a heady period in that quest. The oldest dates rolled back millions of years at a time, from the *Magnolia* leaf (12–17 million years) and the insects in amber (20–30 million years) to the weevil in amber (130 million years). At California Polytechnic State University, Raul Cano was one of the scientists keeping the world abreast of these amazing discoveries. He was as keen as others to get his work published fast, but in late 1991 he encountered something so unusual he decided to remain silent on the matter until it had been checked and double-checked. What he and his student Monica Borucki had noticed concerned an insect in amber from which they were hoping to amplify DNA. The insect's gut contained what looked like bacterial spores. Having sterilized the amber and everything around these spores, they transferred them to a sterile nutrient solution. They then witnessed something that is still hard to believe. A single-cell organism, trapped in amber for 30 million years, seemed to be coming back to life and growing.

Cano and Borucki waited for three years of further tests before publishing their findings in *Science*. DNA sequencing demonstrated that the bacterium was distinct from any modern species, but its closest living relatives were gut flora bacteria. Both these points supported the notion of the spores being from the ancient insect's gut flora. By the time they had the confidence to publish in *Science*, they had sufficient results to embark on exploiting them in a unique way. Cano set up a company, Ambergene, that was devoted to reviving these ancient organisms, and he patented the process of revival. Ambergene has since claimed to have revived species by the thousand, some from over 100 million years of dormancy in amber. Jurassic Park might just be possible for bacteria, if not for dinosaurs.

A large question mark now hangs over the recovery of highly fragmentary DNA within amber-entombed insects. One might imagine that the recovery of viable bacteria within those same insects' bodies was equally problematic. It would involve the survival of complete DNA sequences, not just the tiny fragments normally recovered as ancient DNA. This, however, is not as crazy as it seems. Many who have queried the insect DNA have reserved judgement on the more ambitious claim for bacterial DNA. This is for two reasons. First, the bacterial spore is a remarkable DNA survival capsule, which could well display a molecular resilience not shared by the surrounding host. Second, we are becoming increasingly familiar with other bacterial 'prisons', in which some combination of dormancy and ultra-slow metabolism allows bacteria to persist entrapped for millions of years. Researchers boring into the sea floor have found viable bacteria beneath a kilometre of sediment, and others have found ancient bacteria in the heart of geological rock salt, still viable within tiny saline vacuoles in the salt. These survivals are both well attested, so why not also within an amber-entombed host?

Cano is not alone in claiming to have revived ancient bacteria. From rather younger samples, another group found evidence of bacteria in the gut of one of North America's extinct megafauna, the elephant-like mastodon. Specimens of this beast had been recovered from a couple of late Pleistocene ponds in Ohio and Michigan. Within their massive frames were lumps of what looked like the preserved remnants of their large and small intestines. Here, too, bacteria were encountered that could apparently be brought back to life.

beyond amber

We do not yet know how long these spores can remain viable, but even when they fragment and their contents disappear, there are still biomolecular features from within these least visible components of the living world that outlast all others. Franco Rollo's work has demonstrated bacterial DNA persisting outside the spores for thousands of years, and Angela Gernaey's research has shown that the lipids around the bacterial cell can also persist. Within the lipids of the cell coat in many bacteria can be found one of the most durable biomolecules of all. These belong to a family of molecules known as the hopanes, made up of five interlinked carbon rings with a long-chain strand projecting from them. These chains will fragment with time, but the characteristic set of rings and part of the attached chain can persist indefinitely. They have been found in rocks that are billions of years old, providing evidence of some of the earliest life on earth. These seemingly indestructible hopanoid molecules are widespread as traces within petroleum deposits. Here they retain some of the characteristic structure of the ancient bacteria that millions, or billions, of years ago formed part of the living worlds from which these deposits derive. Ian Head and his colleagues at Newcastle are currently attempting to decode the hopanoid signatures from deep within the Earth's crust, this most enduring legacy of all the world's ancient microbial populations.

These characteristic molecules, whose age is of the same order as the Earth itself, epitomize the transformation underway in the molecular approach to the past. A generation ago, many plants and animals remained archaeologically invisible. We had not yet learnt how to seek out and identify the scattered fragments of tissue they left behind. The seeking out of microbes was far beyond the aspirations of the first generation of bio-archaeologists. The molecule hunt has turned that around. What seemed at first to be the least accessible organisms from the past can now be seen to leave lipid and DNA signatures as clear as anything from a more conspicuous organism. Furthermore, these signatures cling tenaciously, first to life itself, and then to the molecular traces of life, for periods of time during which every other component of life has been broken down and recycled millions of times over.

11

the hunt goes on

a molecular spotlight

The roots of biomolecular archaeology can be traced back many years – at least as far as Lyle Boyd's search for blood groups of ancient mummies in the 1930s. Today, the number of ancient bodies assayed for ancient DNA runs into four figures, and molecular traces have been identified from seeds and animal bones, from inside pots and from the surface of tools of stone. The time has come to ask how many of the early hopes and aspirations have been realized.

What always follows the excitement of the opening up of a new avenue of science is the realization of its pragmatic limitations, and here the limitations are to how much those ancient biomolecules can reveal. However well preserved, they never actually cease breaking down. The more tattered and fragmentary they become, the less easy it is to make sense of them and to deal with such issues as contamination and the demands of fine-resolution analysis. Space, time and context all impose limits on what is possible. Within those limits, we have learnt a great deal about the Neanderthal question, about human migrations and the development of farming, about cooking and about ancient disease. At the same time, these limits still leave many targets beyond our reach, at least for the moment. Many new avenues of scientific research come up against such a boundary. Two other limitations upon the new science are rather different, and reveal something about the limits of what is in principle knowable about the human past.

In several instances, successful biomolecular projects have shone a spotlight on a particular community caught up in one of our grand narratives of the past. These could be narratives about progress, social structure, revolutionary change or population movement, each in some way translatable into patterns of lineage and kinship. Our first expectation is for that spotlight to reveal evidence that either supports or disputes that narrative. What it has revealed instead is that life tends to be more complicated than that. Where one major journey was under the molecular spotlight, a complex of subsidiary journeys began to appear. Where a single domestication was sought, a double or triple domestication was found. The situation seems to me to resemble another, much earlier, transformation in the scale of our observations. When, three centuries ago, van Leeuwenhoek first looked through a microscope at water from a pond, what seemed like clear transparent fluid erupted under his lens into a welter of life on a hitherto unobserved scale. Microscopy has since dramatically demonstrated how patterns that seem to be simple at one level, at a finer scale of resolution resolve into patterns that are more complex. Then, at yet finer levels, they again break up into patterns of greater or lesser complexity. Neither in society nor in nature is the microcosm simply a more detailed snapshot of the macrocosm. In society, the distinctiveness of the microcosm is further enhanced by a particular attribute of being human, the nature of human action.

The sharper the focus of the molecular spotlight upon particular events and individuals in the past, the more directly it will illuminate the freedom of human action. There may be broader patterns, relationships and cultural rules, but we humans by our very nature are constantly reworking patterns, renegotiating relationships and breaking rules. This is where biomolecular applications in archaeology come very close to those in forensic science, which is all about the analysis of broken rules. As we rebuild a particular human episode from the past, with the precision that biomolecular science allows, then the complexity that arises from a constellation of individual lives will be far more evident than it is in the patterns from a distance that inform the grand narratives.

The projects featured in the preceding pages have tended to blur the edges of many of the simpler narratives of the human past, and instead reveal the diversity of the lives that cluster around those narratives. In the case of the origins and spread of agriculture, for example, waves of migrating farmers are gradually being superseded in our accounts by meetings within a diverse world. This is a world in which roots, tubers and leaves are as much a part of the story as cereals. It encompasses regions and time scales lying well outside those enshrined in stories of the Fertile Crescent. At the same time as such patterns of complexity and diversity fall under the molecular spotlight, a beam is also cast upon a new generation of narratives emerging from the molecular sciences themselves.

Biomolecular archaeology came about as a result of two different disciplines moving forward in tandem. Biology was gaining knowledge and control of its own molecular architecture, at a considerable pace. At the same time, archaeology was reaching out to explore a different range of ancient materials beneath the ground. Through luck and circumstance the two converged. It was less a case of dredging up some tried and tested scientific tools for archaeological application, and more the meeting of two independent research frontiers, uncertainly coming together on what emerged as shared questions. This uncertain convergence is still going on, perhaps most obviously in the studies of human genetic diversity. Upon the groundwork laid by Luca Cavalli-Sforza, the mapping of the human genome and its variation is continuing apace. As more and more DNA data from living humans becomes available, it comes rich in hints about early episodes in the human past. Much as the intense mapping of the mitochondrial genome resulted in a new treasure chest of data on the human family tree, so the study of hot spots along the Y chromosome is having a similar effect, and other chromosomes will follow suit. As patterns of variation emerge, so resonances are sought with the various principal components of Cavalli-Sforza's gene maps. In addition, entirely new spatial patterns are coming to light. As a first approximation, the overworked molecular clock places these patterns in some broad time frame, in line with hypotheses about how the human past was shaped. To an archaeologist this rapid expansion of the genetic database is reminiscent of an earlier growth in the evidence of human variation. It is similar to the blossoming of the world-wide archaeological record in the mid-twentieth century. That too depended on an over-stretched clock, derived from sequences of artefact styles. It was rationalized and corrected by much tighter chronological controls from carbon dating and other radiometric chronologies. The same needs to happen with the wealth of new genetic patterns and ancient DNA. Fragments of genetic information from well-dated, secure contexts in the archaeological record are what will anchor the broader modern pattern in time as well as space.

DNA and its fellow travellers

DNA has been the flagship in a fleet of molecules that are moving our understanding of the human past forward. In the future, we are likely to see the different biomolecules brought together to elucidate that past. Three of these molecules are currently being used to investigate the prehistoric origins of dairying – the DNA of the cattle themselves, and the characteristic lipid and protein signatures left behind in the milk pots. DNA and lipid analysis have been brought together in the study of disease. A few ancient cemetery studies are now bringing together DNA analyses and isotopic studies of bone collagen.

We have perceived the limits of looking at ancient DNA alone. If Matthew Collins's thermal projections of burial conditions are right, only nine known Neanderthal individuals are expected to retain sufficient intact DNA in a form that is accessible with our existing techniques. Indeed, they may be the only nine individuals of any hominid species other than our own to do so. This brings our thoughts back to the proteins that played such an important role in helping us construct a picture of human genetic diversity in the first place. There is a reasonable case for the persistence of some recognizable protein fragments in dinosaur bones tens of millions of years old, and in marine invertebrates hundreds of millions of years old. Perhaps we shall before long learn something of the proteins of *Homo erectus* and other key species, and assemble new stories around that. However, the latest suggestions are that the whole biomolecular picture could in any case be changed yet again.

a solid state future

The major enemy of ancient molecules is undoubtedly water. Chemists describe it as the 'universal catalyst', the trigger whose presence is needed for so many of the chemical reactions we encounter, particularly those within living systems. We have seen how the very dry sites have conserved a rich array of biomolecules, and the same is true on a microscopic scale. If a solid tissue such as wood or bone excludes water by its dense internal structure, then biomolecules can persist even when the bone or wood fragment has spent thousands of years immersed in soggy peat. At an even smaller scale, the power of solidity in protecting molecules from water becomes yet more marked. This can be observed as we move from a solid bone to a tiny particle of hydroxy-apatite mineral, or from a piece of wood to phytoliths and phytocrystals. The long-chain molecules that are particularly vulnerable to prolonged contact with water can form weak bonds along the chain's edges. These bonds collectively become strong when they wrap the molecule across the surface of a mineral particle, or become consumed by it. Wholly immersed, or clinging on tightly to these solid state sanctuaries within a world of breakdown and decay, these biomolecules may last indefinitely. The problem is that they are not currently within the range of analytical methods commonly in use. So tightly are they bound, prising them off would probably destroy them anyway. It is the slightly freer, and consequently more vulnerable, molecules that are being detected by present methods. However, a start has now been made in seeking those in hiding in the solid state zone.

At Menez-Dregan in Brittany, a collapsed cave encased a series of hearths and bone fragments left behind after the visits of *Homo erectus* half a million years ago. The bones did not look exceptional, except for the manner of their burial. The cave's collapse had greatly compressed them, and they had rapidly desiccated under high ionic pressure. As the site was excavated, molecular biologist Eva-Maria Geigl came across from Paris to extract the bones carefully and put them in cold storage. Aware of the problems outlined above, Geigl decided to try to detect any surviving DNA *in situ* in the bone tissue. She attempted this by hybridizing DNA probes on to the bone surface, without trying to remove the molecules – and she got a positive result. It appeared that in the right circumstances (and high pressure may be key), the possibility of really ancient DNA might once again be opened up. It is too early to say how well her results will be received, but at least there is not the mismatch between empirical results and theoretical kinetics that there was with the earlier amplifications from dinosaur bone. Whatever the fate of Geigl's results, tightly attached biomolecules that effectively persist in solid state have a promising future.

Something similar is happening within ancient protein work, where again the best survivors are probably clinging on tight to a solid surface. At Newcastle, Matthew Collins's team realized this when they embarked on their search for ancient dairying. They were targeting residues of the milk protein casein in pottery fragments. Not long ago, I heard one of the group, Oliver Craig, give a talk on their ingenious approach to tracking down casein molecules tightly adsorbed on to the pottery surface. They had reasoned that if this tightly bound protein could not be detached intact from the pottery fabric, then the pottery fabric had to be detached from the protein. The method they developed entailed placing the pot fragments in a very particular kind of tube whose inner surface acted like a molecular 'fly-paper' for free proteins. They then set about dissolving the pottery by adding the powerful solvent hydrofluoric acid to the tube. As soon as any surviving proteins were free of their pottery attachments, they clung to their new home on the side of the tube. Here, they could be identified by monoclonal antibodies.

As I listened, fascinated, my mind drifted back to that first experience of practical archaeology in a Somerset field in the 1960s. What Oliver was describing was a complete inversion of our data collection then. We had sat between the excavation trench and the site hut, nailbrushes in hand, scrubbing away the dirty, smelly biomolecules from the inert sherds of pottery. At Newcastle, thirty years or so later, the priorities had completely changed. It was now the pottery that was being chemically scrubbed away from the biomolecules. It was a passing thought, and I do not intend to suggest that our entire ceramic heritage should be plunged into a vat of hydrofluoric acid. But looking back on that prehistoric village, I am struck by how much our perception of what remains of the past has changed. In many ways, the visibility and the durability of the finds has proven a poor reflection of the range of information accessible by first bio-archaeological and then biomolecular means.

When David Clarke reconsidered the renowned prehistoric lake villages of Somerset just over twenty-five years ago, he set his sights beyond pottery typologies and metalwork affinities to what life was actually like in the first millennium BC. He speculated about a community adapting to a watery world in which the water was both friend and foe. It was friend in providing a host of resources – shellfish, water plants and rhizomes, nuts, fuels, reeds and sedges. It was a foe in harbouring foot rot, parasites and insect-borne diseases, and forcing the farmers to move their herds around the landscape to graze on different pastures in different seasons. He considered how these lives were put together and how the village communities connected with their neighbours, but he had very little evidence with which to explore any of his suppositions. In the following years, the bio-archaeologists working in John Coles's Somerset Levels Project would provide some of that evidence, in the form of organic remains that could be studied either with the naked eye or under an optical microscope. Today, virtually every avenue of his speculations could be followed through biomolecules. DNA could reveal much about the local human community, its pattern of kinship, and its interrelations with its more distant neighbours. The same molecule could probe the water-borne diseases suffered by both the farming families and their herds. Isotopic studies of bone proteins could track the dietary journey of those travelling herds, and piece together the diet of their human owners. Lipid studies could be targeted upon that long list of plant and animal resources, many of which fail to show up in any other way. These are possibilities yet to be realized, but the scientific foundation for their realization is in place.

references

1 a different kind of past

Ancient Biomolecules (1999) 2:2–3, Special Issue, *The* Ancient Biomolecules *Initiative*.

Boyd, W. C. and Boyd, L. G. (1933) 'Blood grouping by means of preserved muscle', *Science* 78: 578.

Bulleid, A. (1926) *The Lake-villages of Somerset* (2nd edition), London: Folk Press.

Clarke, D. L. (1972) 'A provisional model of an Iron Age society and its settlement system', in D. L. Clarke (ed.), *Models in Archaeology*, London: Methuen, pp. 801–69.

Coles, J. M. (1987) *Meare Village East: The Excavations of A. Bulleid and H. St George Gray 1932–1956*, Exeter: Somerset Levels Project.

Coles, J. M. and Coles, B. (1986) *Sweet Track to Glastonbury: The Somerset Levels in Prehistory*, London: Thames and Hudson.

Cunliffe, B. W. and Miles, D. (eds) (1984) *Aspects of the Iron Age in Central Southern Britain*, Oxford University: Committee for Archaeology.

Eglinton, G. and Curry, G. B. (eds) (1991) *Molecules through Time: Fossil Molecules and Biochemical Systematics*, London: Royal Society.

Evans, J. and Hill, H. E. (1982) 'Dietetic information by chemical analysis of Danish Neolithic pot sherds: a progress report', in A. Aspinall and S. E. Warren (eds), *Proceedings of the 22nd Symposium on Archaeometry*, University of Bradford, pp. 224–8.

Gould, S. J. (1989) *Wonderful Life: The Burgess Shale and the Nature of History*, London: Penguin.

Higuchi, R. G. *et al.* (1984) 'DNA sequences from the quagga, an extinct member of the horse family', *Nature* 312: 282–4.

Higuchi, R. G. *et al.* (1987) 'Mitochondrial DNA of the extinct quagga: relatedness and extent of post-mortem change', *Journal of Molecular Evolution* 25: 283–7.

Rottländer, R. C. A. and Schlichterle, H. (1979) 'Food identification of samples from archaeological sites', *Archaeophysika* 10: 260–7.

Turner, R. C. and Scaife, R. G. (1995) *Bog Bodies: New Discoveries and New Perspectives*, London: British Museum Press.

Wang, G. H. and Lu, C. L. (1981) 'Isolation and identification of nucleic acids of the liver from a corpse from the Changssha Han tomb', *Shen Wu Hua Hsueh Yu Sheng Wu Li Chin Chan* 17: 70–5.

2 the quest for ancient DNA

Aldhous, P. (1996) 'Dinosaur DNA fails new test of time', *New Scientist* 150 (2030): 21.

Allard, M. W., Young, D. and Huyen, Y. (1995) 'Detecting Dinosaur DNA', *Science* 268: 1192.

Austin, J. J. *et al.* (1997) 'Problems of reproducibility – does geologically ancient DNA survive in amber-preserved insects?' *Proceedings of the Royal Society of London*, Series B, 264: 467–74.

Austin, J. J. *et al.* (1998) 'Ancient DNA from amber inclusions: a review of the evidence', *Ancient Biomolecules* 2 (2): 167–76.

Bada, J. L. *et al.* (1994) 'Amino acid racemization in amber-entombed insects: implications for DNA preservation', *Geochimica et Cosmochimica Acta* 58 (14): 3131–5.

Brown, T. A. and Brown, K. A. (1992) 'Ancient DNA and the archaeologist', *Antiquity* 66: 10–23.

Brown, T. A. *et al.* (1993) 'Biomolecular archaeology of wheat: past, present and future', *World Archaeology* 25: 64–73.

Brown, T. A. *et al.* (1994) 'DNA in wheat seeds from European archaeological sites', *Experientia* 50 (6): 571–5.

Cano, R. J., Poinar, H. and Poinar, G. (1992) 'Isolation and partial characterisation of DNA from the bee *Proplebeia dominicana* (Apidae: Hymenoptera) in 25–40-million-year-old amber', *Medical Science Research* 20: 249–51.

Cano, R. J. *et al.* (1992) 'Enzymatic amplification and nucleotide sequencing of portions of the 18s rRNA gene of the bee *Proplebeia dominicana* (Apidae: Hymenoptera) isolated from 25–40-million-year-old Dominican amber', *Medical Science Research* 20: 619–22.

Cano, R. J. *et al.* (1993) 'Amplification and sequencing of DNA from a 120–135-million-year-old weevil', *Nature* 363: 536–8.

DeSalle, R., Barcia, M. and Wray, C. (1993) 'PCR jumping in clones of 30-million-year-old DNA fragments from amber-preserved termites (*Mastotermes electrodominicus*)', *Experientia* 49: 906–9.

Desalle, R. *et al.* (1992) 'DNA sequences from a fossil termite in Oligo-Miocene amber and their phylogenetic implications', *Science* 257: 1933–6.

Doran, G. H. *et al.* (1986) 'Anatomical, cellular and molecular analysis of 8,000-yr-old human brain tissue from the Windover archaeological site', *Nature* 323: 803–6.

Golenberg, E. M. (1991) 'Amplification and analysis of Miocene plant fossil DNA',

Philosophical Transactions of the Royal Society of London, Series B, 333: 419–27.

— (1994) 'DNA from plant compression fossils. Ancient DNA', in B. Herrmann and S. Hummel (eds), *Ancient DNA. Recovery and Analysis of Genetic Material from Palaeontological, Archaeological, Museum, Medical and Forensic Specimens*, New York: Springer-Verlag, pp. 237–56.

Golenberg, E. M. *et al.* (1990) 'Chloroplast DNA sequence from a Miocene *Magnolia* species', *Nature* 344: 656–8.

Hagelberg, E. and Clegg, J. B. (1991) 'Isolation and characterization of DNA from archaeological bone', *Proceedings of the Royal Society of London*, Series B, 244: 45–50.

Hagelberg, E., Sykes, B. C. and Hedges, R. E. M. (1989) 'Ancient bone DNA amplified', *Nature* 342: 485.

Hedges, R. E. M. and Sykes, B. C. (1991) 'Biomolecular archaeology: past, present and future', in M. Polland (ed.), *New Developments in Archaeological Science*, London: Oxford University Press, pp. 267–83.

Hedges, S. B. and Schweitzer, M. H. (1995) 'Detecting dinosaur DNA', *Science* 268: 1191–2.

Higuchi, R. G. and Wilson, A. C. (1984) 'Recovery of DNA from extinct species', *Federation Proceedings. (Federation of American Societies for Experimental Biology)* 43: 1557.

Higuchi, R. G. *et al.* (1984) 'DNA sequences from the quagga, an extinct member of the horse family', *Nature* 312: 282–4.

Higuchi, R. G. *et al.* (1987) 'Mitochondrial DNA of the extinct quagga: relatedness and extent of post-mortem change', *Journal of Molecular Evolution* 25: 283–7.

Lindahl, T. (1993) 'Instability and decay of the primary structure of DNA', *Nature* 362: 709–15.

— (1993) 'Recovery of antediluvian DNA', *Nature* 365: 700.

— (1995) 'Recognition and processing of damaged DNA', *Journal of Cell Science* 49: 73–7.

— (1997) 'Facts and artifacts of ancient DNA', *Cell* 90: 1–3.

Logan, G. A., Boon, J. J. and Eglinton, G. (1993) 'Structural biopolymer preservation in Miocene leaf fossils from the Clarkia site, northern Idaho', *Proceedings of the National Academy of Science USA* 90: 2246–50.

Mullis, K. B. and Faloona, F. A. (1987) 'Specific synthesis of DNA in vitro via a polymerase catalysed chain reaction', *Methods in Enzymology* 155: 335–50.

Pääbo, S. (1985) 'Molecular cloning of ancient Egyptian mummy DNA', *Nature* 314: 644–5.

— (1985) 'Preservation of DNA in ancient Egyptian mummies', *Journal of Archaeological Science* 12: 411–17.

— (1987) 'Molecular genetic methods in archaeology: a prospect', *Anthropologischer Anzeiger* 45: 9–17.

— (1989) 'Ancient DNA: extraction, characterization, molecular cloning, and enzymatic amplification', *Proceedings of the National Academy of Science USA* 86: 1939–43.

Pääbo, S. and Wilson, A. C. (1988) 'Polymerase chain reaction reveals cloning artefacts', *Nature* 334: 387–8.

— and — (1991) 'Miocene DNA sequences – a dream come true?' *Molecular Evolution* 1 (1): 45–6.

Pääbo, S., Gifford, J. A. and Wilson, A. C. (1988) 'Mitochondrial DNA sequences from a 7000-year old brain', *Nucleic Acids Research* 16 (20): 9775–83.

Pääbo, S., Higuchi, R. and Wilson, A. C. (1989) 'Ancient DNA and the polymerase chain reaction', *Journal of Biological Chemistry* 264 (17): 9709–12.

Poinar, G. O. (1994) 'The range of life in amber: significance and implications in DNA studies', *Experientia* 50 (6): 536–42.

Poinar, G. O. and Hess, R. (1982) 'Ultrastructure of 40-million-year-old insect tissue', *Science* 215: 1241–2.

Poinar, G. O., Poinar, H. N. and Cano, R. J. (1994) 'DNA from amber inclusions', in B. Herrmann and S. Hummel (eds), *Ancient DNA. Recovery and Analysis of Genetic Material from Palaeontological, Archaeological, Museum, Medical and Forensic Specimens*, New York: Springer-Verlag, pp. 92–103.

Poinar, H. N. and Stankiewicz, B. A. (1999) 'Protein preservation and DNA retrieval from ancient tissues', *Proceedings of the National Academy of Science USA* 96: 8426–31.

Poinar, H. N., Cano, R. J. and Poinar, G. O. (1993) 'DNA from an extinct plant', *Nature* 363: 677.

Poinar, H. N. *et al.* (1996) 'Amino acid racemization and the preservation of ancient DNA', *Science* 272: 864–6.

Sidow, A., Wilson, A. C. and Pääbo, S. (1991) 'Bacterial DNA in Clarkia fossils', *Philosophical Transactions of the Royal Society of London*, Series B, 333: 429–33.

Soltis, S., Soltis, D. E. and Smiley, C. J. (1992) 'An *rbc*L sequence from a Miocene *Taxodium* (bald cypress)', *Proceedings of the National Academy of Science USA* 89: 449–51.

Tuross, N. (1994) 'The biochemistry of ancient DNA in bone', *Experientia* 50: 530–5.

Wang, G. H. and Lu, C. L. (1981) 'Isolation and identification of nucleic acids of the liver from a corpse from the Changssha Han tomb', *Shen Wu Hua Hsueh Yu Sheng Wu Li Chin Chan* 17: 70–5.

Watson, J. D. and Crick, F. C. (1953) 'Molecular structure of nucleic acids: a structure for deoxyribose nucleic acids', *Nature* 171: 737–8.

Woodward, S. R., Weyand, N. J. and Bunnell, M. (1994) 'DNA sequence from Cretaceous Period bone fragments', *Science* 266: 1229–32.

3 our curious cousins

Anderson, S. *et al.* (1981) 'Sequence organization of the human mitochondrial genome', *Nature* 290: 457–64.

Cann, R. L. (1992) 'The search for Eve', *Science* 256: 79.

Cann, R. L., Stoneking, M. and Wilson, A. C. (1987) 'Mitochondrial DNA and human evolution', *Nature* 325: 31–6.

Cooper, A. *et al.* (1997) 'Neandertal genetics', *Science* 277: 1021–4.

Darwin, C. R. (1859) *The Origin of Species by Means of Natural Selection, or the Preservation of Favoured Races in the Struggle for Life*, London: John Murray.

Horai, S. (1995) 'Evolution and the origins of man: clues from complete sequences of hominoid mitochondrial DNA', *Southeast Asian Journal of Tropical Medicine and Public Health* 26 (suppl. 1): 146–54.

Horai, S. *et al.* (1995) 'Recent African origin of modern humans revealed by complete sequences of hominoid mitochondrial DNA', *Proceedings of the National Academy of Science USA* 92: 532–6.

Huxley, T. H. (1863) *Man's Place in Nature*, London: Williams and Norgate.

King, W. (1864) 'The reputed fossil man of Neandertal', *Quarterly Journal of Science* I: 88–97.

Krings, M. *et al.* (1997) 'Neandertal DNA sequences and the origin of modern humans', *Cell* 90: 19–30.

Krings, M. *et al.* (1999) 'DNA sequence of the mitochondrial hypervariable region II from the Neandertal type specimen', *Proceedings of the National Academy of Science USA* 96: 5581–5.

Krings, M. *et al.* (2000) 'A view of Neandertal genetic diversity', *Nature Genetics* 26: 144–6.

Ovchinnikov, I. V. *et al.* (2000) 'Molecular analysis of Neanderthal DNA from the northern Caucasus', *Nature* 404: 490–3.

Sarich, V. M. and Wilson, A. C. (1967) 'Immunological time scale for hominid evolution', *Science* 158: 1200–03.

Stringer, C. B. (1990) 'The emergence of modern humans', *Scientific American* 263 (6): 98–104.

Stringer, C. B. and Andrews, P. (1988) 'Genetic and fossil evidence for the origin of modern humans', *Science* 239: 1263–8.

Stringer, C. B. and McKie, R. (1996) *African Exodus: The Origins of Modern Humans*, London: Jonathan Cape.

Tattersall, I. and Schwartz, J. H. (1999) 'Hominids and hybrids: the place of Neanderthals in human evolution', *Proceedings of the National Academy of Science USA* 96: 7117–19.

Thorne, A. and Wolpoff, M. (1991) 'Conflict over modern human origins', *Search* 22: 175–7.

— and — (1992) 'The multiregional evolution of humans', *Scientific American* (April): 28–33.

Wolpoff, M. and Thorne, A. (1991) 'The case against Eve', *New Scientist* 130: 37–41.

Wolpoff, M. H., Wu, X. and Thorne, A. (1984) 'Modern *Homo sapiens* origins: a general theory of hominid evolution involving the fossil evidence from East Asia', in F. Smith and F. Spencer (eds), *The Origins of Modern Humans: A World Survey of the Fossil Evidence*, New York: Alan Liss, pp. 411–83.

Zuckerkandl, E. and Pauling, L. (1965) 'Evolutionary divergence and convergence in proteins', in V. Bryson and H. J. Vogel (eds), *Evolving Genes and Proteins*, New York: Academic Press, pp. 97–166.

4 final traces of life

Allison, A. (1988) 'The role of anoxia in the decay and mineralization of proteinaceous macro-fossils', *Paleobiology* 14: 139–54.

Arriaza, B. T. (1995) *The Chinchorro Mummies of Ancient Chile*, Washington: Smithsonian.

Bahn, G. (ed.) (1996) *Tombs, Graves and Mummies*, London: Weidenfeld and Nicolson.

Briggs, D. E. G. (1999) 'Molecular taphonomy of animal and plant cuticles: selective preservation and diagenesis', *Philosophical Transactions of the Royal Society of London*, Series B, 354: 7–17.

Briggs, D. E. G., Evershed, R. and Stankiewicz, B. A. (1999) 'The molecular preservation of fossil arthropod cuticles', *Ancient Biomolecules* 2 (2): 135–46.

Cockburn, A. and Cockburn, C. (eds) (1980) *Mummies, Disease, and Ancient Cultures*, Cambridge: Cambridge University Press.

Coles, J. M. (1984) *The Archaeology of Wetlands*, Edinburgh: Edinburgh University Press.

Collins, M. J., Waite, E. R. and Van Duin, A. C. T. (1999) 'Predicting protein decomposition: the case of aspartic-acid racemization kinetics', *Philosophical Transactions of the Royal Society of London*, Series B, 354: 51–64.

Collinson, M. E. *et al.* (1999) 'The preservation of plant cuticle in the fossil record: a chemical and microscopical investigation', *Ancient Biomolecules* 2 (2): 251–65.

Eglinton, G. and Logan, G. A. (1991) 'Molecular preservation', *Philosophical Transactions of the Royal Society of London*, Series B, 333: 315–28.

Gay, S. and Miller, E. J. (1978) *Collagen in the Physiology and Pathology of Connective Tissue*, New York: Fisher.

Gemmel, R. T., McGerity, T. J. and Grant, W. D. (1999) 'Use of molecular techniques to investigate possible long-term dormancy of halobacteria in ancient halite deposits', *Ancient Biomolecules* 2 (2): 125–33.

Hansen, J. P. H. (1991) *The Greenland Mummies*, London: British Museum Publications.

Painter, T. J. (1991) 'Lindow Man, Tollund Man and other peat-bog bodies: the preservative and antimicrobial action of sphagnan, a reactive glycuronoglycan with tanning and sequestering properties', *Carbohydrate Polymers* 15: 123–42.

Parkes, R. J. *et al.* (1997) 'A deep bacterial biosphere in Pacific Ocean sediments', *Nature* 371: 410–13.

Price, T. D. (ed.) (1989) *The Chemistry of Prehistoric Human Bone*, Cambridge: Cambridge University Press.

Reynolds, T. M. (1965) 'Chemistry of non-enzymic browning. II', *Advances in Food Research* 14: 167–283.

Turner, R. C. and Scaife, R. G. (1995) *Bog Bodies: New Discoveries and New Perspectives*, London: British Museum Press.

Uhle, M. (1917) 'Los aborigenes de Arica', *Publicaciones del Museo de Etnologia y Antropologia de Chile*, Santiago, Chile 1 (4–5): 151–76.

5 gaining control

Allaby, R. G., Banerjee, M. and Brown, T. A. (1999) 'Evolution of the high-molecular-weight glutenin loci of the A, B, D and G genomes of wheat', *Genome* 42: 296–307.

Allaby, R. G., Jones, M. K. and Brown, T. A. (1994) 'DNA in charred wheat grains from the Iron Age hillfort at Danebury, England', *Antiquity* 68 (258): 126–32.

Allaby, R. G. *et al.* (1997) 'Evidence for the survival of ancient DNA in charred wheat seeds from European archaeological sites', *Ancient Biomolecules* 1 (2): 119–29.

Barker, G. (1985) *Prehistoric Farming in Europe*, Cambridge: Cambridge University Press.

Bar-Yosef, O. and Belfer-Cohen, A. (1989) 'The origins of sedentism and farming communities in the Levant', *Journal of World Prehistory* 3: 477–98.

Bar-Yosef, O. and Gopher, A. (eds) (1997) 'An early Neolithic village in the Jordan Valley I. The archaeology of Netiv Hagdud', *American School of Prehistoric Research Bulletin* 43, Cambridge: Peabody Museum of Archaeology and Ethnology, pp. 247–66.

Braidwood, L. *et al.* (eds) (1983) *Prehistoric Archaeology along the Zagros Flanks*, Chicago: University of Chicago Oriental Institute Publications v105.

Brown, T. A. (1999) 'How ancient DNA may help in understanding the origin and spread of agriculture', *Philosophical Transactions of the Royal Society of London*, Series B, 354: 89–98.

Brown, T. A. and Brown, K. A. (1992) 'Ancient DNA and the archaeologist', *Antiquity* 66 (250): 10–23.

Brown, T. A. *et al.* (1993) 'Biomolecular archaeology of wheat: past, present and future', *World Archaeology* 25: 64–73.

Brown, T. A. *et al.* (1994) 'DNA in wheat seeds from European archaeological sites', *Experientia* 50 (6): 571–5.

Brown, T. A. *et al.* (1999) 'Ancient DNA in charred wheats: taxonomic identification of mixed and single grains', *Ancient Biomolecules* 2: 185–93.

Chen, W. B. (1993) 'Indica-Japonica differentiation and its relevance to the domestication process in rice: bioarchaeological and molecular genetic studies', Doctoral dissertation, Gifu University, Gifu, Japan.

Chen, W. B. *et al.* (1993) 'Distribution of deletion type in CpDNA of cultivated and wild rice', *Japanese Journal of Genetics* 68: 597–603.

Childe, V. G. (1936) *Man Makes Himself*, London: Watts and Co.

— (1942) *What Happened in History*, New York: Penguin Books.

Darwin, C. R. (1859) *The Origin of Species by Means of Natural Selection, or the Preservation of Favoured Races in the Struggle for Life*, London: John Murray.

— (1868) *The Variation of Animals and Plants under Domestication*, Volume 1. London: John Murray.

de Candolle, A. (1882) *Origine des plants cultivées*, Paris: Germer Baillière.

Deakin, W. J., Rowley-Conwy, P. and Shaw, C. H. (1998) 'Amplification and sequencing of DNA from preserved sorghum of up to 2,800 years antiquity found at Qasr Ibrim', *Ancient Biomolecules* 2 (1): 27–41.

—, — and — (1999) 'The sorghum of Qasr Ibrim: reconstructing DNA templates from ancient seeds', *Ancient Biomolecules* 2 (2): 117–24.

Doebley, J. F. (1990) 'Molecular evidence and the evolution of maize', in K. Bretting (ed.), 'New perspectives on the origin and evolution of New World domesticated plants', *Economic Botany* 44: 6–28.

— (1992) 'Molecular systematics and crop evolution', in S. Soltis, D. E. Soltis and J. J. Doyle (eds), *Molecular Systematics of Plants*, New York/London: Chapman and Hall, pp. 202–22.

— (1995) 'Genetics, development, and the morphological evolution of maize', in C. Hock and A. G. Stephenson (eds), *Experimental and Molecular Approaches to Plant Biosystematics*, St Louis: Missouri Botanic Gardens, pp. 57–70.

Doebley, J. F., Goodman, M. M. and Stuber, C. W. (1984) 'Isozymic variation in *Zea*', *Systematic Botany* 9: 203–18.

Goloubonoff, P., Pääbo, S. and Wilson, A. C. (1993) 'Evolution of maize inferred from sequence diversity of an Adh2 gene segment from archaeological specimens', *Proceedings of the National Academy of Science USA* 90: 1997–2001.

Harris, D. R. (1990) 'Vavilov's concept of centres of origin of cultivated plants: its genesis and its influence on the study of agricultural origins', *Biological Journal of the Linnean Society* 39: 7–16.

Harris, D. R. and Hillman, G. C. (1989) *Foraging and Farming: The Evolution of Plant Exploitation*, London: Unwin Hyman.

Heun, M. *et al.* (1997), 'Site of einkorn wheat domestication identified by DNA fingerprinting', *Science* 278: 1312–14.

Higgs, E. S. (ed.) (1972) *Papers in Economic Prehistory*, Cambridge: Cambridge University Press.

Jones, M. K. and Brown, T. (2000) 'Agricultural origins: the evidence of modern and ancient DNA', *Holocene* 10 (6): 775–82.

Jones, M. K., Allaby, R. G. and Brown, T. A. (1998) 'Wheat domestication', *Science* 279: 302–3.

Jones, M. K., Brown, T. A. and Allaby, R. G. (1996) 'Tracking early crops and early farmers: the potential of biomolecular archaeology', in D. R. Harris (ed.), *The Origins and Spread of Agriculture and Pastoralism in Eurasia*, London: University College London, pp. 93–100.

Kislev, M. E., Nadel, D. and Carmi, I. (1992) 'Epipalaeolithic (19,000 BP) cereal and fruit diet at Ohalo II, Sea of Galilee, Israel', *Review of Palynology* 73: 161–6.

Lee, R. B. and DeVore, I. (eds) (1968) *Man the Hunter*, Chicago: Aldine.

MacNeish, R. (1967) *Prehistory of the Tehuacan Valley*, Austin: University of Texas Press.

Nakamura, I. and Sato, Y. I. (1991) 'Amplification of chloroplast DNA fragment from a single ancient rice seed', *Rice Genetics* (IRRI Manila, Philippines) 2: 802–5.

— and — (1991) 'Amplification of DNA fragments isolated from a single seed of ancient rice (AD 800) by polymerase chain reaction', *Chinese Journal of Rice Science* 5: 175–9.

O'Donoghue, K. *et al.* (1996) 'Remarkable preservation of biomolecules in ancient radish seeds', *Philosophical Transactions of the Royal Society of London*, Series B, 263: 541–7.

Renfrew, J. M. (1973) *Palaeoethnobotany. The Prehistoric Food Plants of the Near East and Europe*, London: Methuen.

Rogers, S. O. and Bendich, A. J. (1985) 'Extraction of DNA from milligram amounts of fresh herbarium and mummified plant tissues', *Plant Molecular Biology* 5: 69–76.

Rollo, F. (1985) 'Characterisation by molecular hybridisation of RNA fragments isolated from ancient (1400 BC) seeds', *Theoretical and Applied Genetics* 71: 330–3.

Sato, Y. I. (1990) 'Non-random association of genes and characters found in *Indica x Japonica* hybrids of rice', *Heredity* 65: 75–9.

— (1997) 'Cultivated rice was born in the lower and middle basins of the Yangtze River', *Nikkei Science* 1: 32–42.

Vavilov, N. (1992) *The Origin and Geography of Cultivated Plants*, Cambridge: Cambridge University Press.

Zohary, D. (1969) 'The progenitors of wheat and barley in relation to domestication and agricultural dispersal in the Old World', in P. J. Ucko and G. W. Dimbleby (eds), *The Domestication and Exploitation of Plants and Animals*, London: Duckworth, pp. 47–66.

— (1996) 'The mode of domestication of the founder crops of Southwest Asian agriculture', in D. R. Harris (ed.), *The Origins and Spread of Agriculture and Pastoralism in Eurasia*, London: University College London, pp. 142–58.

Zohary, D. and Hopf, H. (1993) *Domestication of Plants in the Old World. The Origin and Spread of Cultivated Plants in West Asia, Europe and the Nile Valley*, Oxford: Oxford University Press.

6 ending the chase

Bailey, J. F. *et al.* (1996) 'Ancient DNA suggests a recent expansion of European cattle from a diverse wild progenitor species', *Proceedings of the Royal Society of London*, Series B, 263: 1467–73.

Bar-Yosef, O. and Kislev, M. E. (1989) 'Early farming communities in the Jordan Valley', in D. R. Harris and G. Hillman (eds), *Foraging and Farming*, London: Unwin Hyman, pp. 632–42.

Bradley, D. G. *et al.* (1996) 'Mitochondrial diversity and the origins of African and European cattle', *Proceedings of the National Academy of Science USA* 93: 5131–5.

Bradley, D. G. *et al.* (1998) 'Genetics and domestic cattle origins', *Evolutionary Anthropology* 6 (3): 79–86.

Cooper, A. *et al.* (1992) 'Independent origins of New Zealand moas and kiwis', *Proceedings of the National Academy of Science USA* 89: 8741–4.

Derenko, M., Malyarchuk, B. and Shields, G. F. (1997) 'Mitochondrial cytochrome b sequence from a 33000 year-old woolly mammoth (*Mammuthus primigenius*)', *Ancient Biomolecules* 1 (2): 149–53.

Gimbutas, M. (1952) 'On the origin of north Indo-Europeans', *American Anthropologist* 54 (4): 602–11.

— (1970) 'Proto-Indo-European culture: the Kurgan culture during the fifth to third millennia BC', in G. Cardona, H. M. Koeningswald and A. Senn (eds), *Indo-European and Indo-Europeans*, Philadelphia: University of Pennsylvania Press, pp. 155–98.

Hagelberg, E. *et al.* (1994) 'DNA from ancient mammoth bones', *Nature* 370: 333–4.

Hauf, J. *et al.* (1995) 'Selective amplification of a mammoth mitochondrial cytochrome b fragment using an elephant-specific primer', *Current Genetics* 37 (5): 486–7.

Higgs, E. S. (ed.) (1972) *Papers in Economic Prehistory: Studies by Members and Associates of the British Academy Major Research Project in the Early History of Agriculture*, Cambridge: Cambridge University Press.

Hoss, M., Pääbo, S. and Vereshchagin, N. K. (1994) 'Mammoth DNA sequences', *Nature* 370: 333.

Hoss, M. *et al.* (1996) 'Molecular phylogeny of the extinct ground sloth *Mylodon darwinii*', *Proceedings of the National Academy of Science USA* 93: 181–5.

Ishida, N. *et al.* (1995) 'Mitochondrial DNA sequences of various species of the genus *Equus* with special reference to the phylogenetic relationship between Przewalski's wild horse and domestic horse', *Journal of Molecular Evolution*, 41: 180–8.

Krajewski, C., Buckley, L. and Westerman, M. (1997) 'DNA phylogeny of the marsupial wolf resolved', *Proceedings of the Royal Society of London*, Series B, 264: 911–17.

Krajewski, C. *et al.* (1992) 'Phylogenetic relationships of the thylacine (Mammalia: Thylacinidae) among dasyuroid marsupials: evidence from cytochrome b DNA sequences', *Proceedings of the Royal Society of London*, Series B, 250: 19–27.

Levine, M. *et al.* (1999) *Late Prehistoric Exploitation of the Eurasian Steppe*, Cambridge: McDonald Institute Monographs.

Lister, A. M. *et al.* (1999) 'Ancient and modern DNA in a study of horse domestication', *Ancient Biomolecules* 2 (2): 267–80.

Loftus, R. T. *et al.* (1994) 'Evidence for two independent domestications of cattle', *Proceedings of the National Academy of Science USA* 91: 2757–61.

Marshall, F. (1990) 'Cattle herds and caprine flocks', in P. Robertshaw (ed.), *Early Pastoralists of South-western Kenya*, vol. II, *Memoirs of the British Institute in Eastern Africa*, Nairobi: British Institute in Eastern Africa.

Meadow, R. H. (1996) 'The origins and spread of agriculture and pastoralism in northwestern South Asia', in D. R. Harris (ed.), *The Origins and Spread of Agriculture and Pastoralism in Eurasia*, London: UCL Press.

Noro, M. *et al.* (1998) 'Molecular phylogenetic inference of the woolly mammoth *Mammuthus primigenius*, based on complete sequences of mitochondrial cytochrome b and 12S ribosomal RNA genes', *Journal of Molecular Evolution* 46: 314–26.

Ozawa, T., Hayashi, S. and Mikhelson, V. (1997) 'Phylogenetic position of mammoth and Steller's sea cow within tethytheria demonstrated by mitochondrial DNA sequences', *Journal of Molecular Evolution* 44: 406–13.

Poinar, H. N. *et al.* (1998) 'Molecular coproscopy: dung and diet of the extinct ground sloth *Nothrotheriops shastensis*', *Science* 281: 402–6.

Shoshani, J. *et al.* (1985) 'Proboscidean origins of mastodon and woolly mammoth demonstrated immunologically', *Paleobiology* 11 (4): 429–37.

Stanley, H. F., Kadwell, M. and Wheeler, J. C. (1994) 'Molecular evolution of the family Camelidae – a mitochondrial DNA study', *Proceedings of the Royal Society of London*, Series B, 256: 1–6.

Thomas, R. H., Pääbo, S. and Wilson, A. C. (1990) 'Chance marsupial relationships', *Nature* 345: 393–4.

Thomas, R. H. *et al.* (1989) 'DNA phylogeny of the extinct marsupial wolf', *Nature* 340: 465–7.

Vilà, C., Maldonado, J. E. and Wayne, R. K. (1999) 'Phylogenetic relationships, evolution, and genetic diversity of the domestic dog', *Journal of Heredity* 90 (1): 71–7.

Yang, H., Golenberg, E. and Shoshani, J. (1996) 'Phylogenetic resolution within the

Elephantidae using fossil DNA sequence from the American mastodon (*Mammut americanum*) as an outgroup', *Proceedings of the National Academy of Science USA* 93: 1190–4.

7 great journeys

Ammerman, A. J. and Cavalli-Sforza, L. L. (1973) 'A population model for the diffusion of early farming in Europe', in A. C. Renfrew (ed.), *The Explanation of Culture Change: Models in Prehistory*, London: Duckworth, pp. 343–57.

Arriaza, B. T. (1995) *The Chinchorro Mummies of Ancient Chile*, Washington: Smithsonian.

Bailliet, G. *et al.* (1994) 'Founder mitochondrial haplotypes in Amerindian populations', *American Journal of Human Genetics* 54: 27–33.

Bellwood, P. (1978) *Man's Conquest of the Pacific*, London: Collins.

Bernal, M. (1987) *Black Athena: the Afroasiatic Roots of Classical Civilization* Vol. 1: *The Fabrication of Ancient Greece, 1785–1985*, London: Free Association Books.

Bonatto, S. L. and Salzano, F. M. (1997) 'Diversity and age of the four major mtDNA haplogroups, and their implications for the peopling of the New World', *American Journal of Human Genetics* 61: 1413–23.

Brown, M. D. *et al.* (1998) 'mtDNA haplogroup X: an ancient link between Europe/ Western Asia and North America', *American Journal of Human Genetics* 63: 1852–61.

Cavalli-Sforza, L. L. (1997) 'Genes, peoples, and languages', *Proceedings of the National Academy of Science USA* 94: 7719–24.

Cavalli-Sforza, L. L. and Minch, E. (1997) 'Paleolithic and Neolithic lineages in the European mitochondrial gene pool', *American Journal of Human Genetics* 61: 247–54.

Cavalli-Sforza, L. L. and Piazza, A. (1993) 'Human genomic diversity in Europe: a summary of recent research and prospects for the future', *European Journal of Human Genetics* 1 (1): 3–18.

Cavalli-Sforza, L. L., Menozzi, P. and Piazza, A. (1993) 'Demic expansions and human evolution', *Science* 259: 639–46.

—, — and — (1994) *The History and Geography of Human Genes*, Princeton: Princeton University Press.

Cavalli-Sforza, L. L., Minch, E. and Mountain, J. L. (1992) 'Coevolution of genes and languages revisited', *Proceedings of the National Academy of Science USA* 89: 5620–4, and reply by M. Richards *et al.*, pp. 251–4.

Cavalli-Sforza, L. L. *et al.* (1988) 'Reconstruction of human evolution: bringing together genetic, archaeological, and linguistic data', *Proceedings of the National Academy of Science USA* 85: 6002–6.

Chikhi, L. *et al.* (1998) 'Clines of nuclear DNA markers suggest a largely neolithic ancestry of the European gene pool', *Proceedings of the National Academy of Science USA* 95: 9053–8.

Childe, V. G. (1950) *Prehistoric Migrations in Europe*, London: Kegan Paul.

Easton, R. D. *et al.* (1996) 'mtDNA variation in the Yanomami: evidence for additional New World founding lineages', *American Journal of Human Genetics* 59: 213–25.

Forster, P. *et al.* (1996) 'Origin and evolution of Native American mtDNA variation: a reappraisal', *American Journal of Human Genetics* 59: 935–45.

Gimbutas, M. (1952) 'On the origin of north Indo-Europeans', *American Anthropologist* 54 (4): 602–11.

— (1970) 'Proto-Indo-European culture: the Kurgan culture during the fifth to third millennia BC', in G. Cardona, H. M. Koeningswald and A. Senn (eds), *Indo-European and Indo-Europeans*, Philadelphia: University of Pennsylvania Press, pp. 155–98.

Greenberg, J. H. (1996) 'The "Greenberg" hypothesis', *Science* 274: 1447; discussion, 1448.

Greenberg, J. H., Turner II, C. G. and Zegura, S. L. (1986) 'The settlement of the Americas: a comparison of the linguistic, dental and genetic evidence', *Current Anthropology* 27: 477–97.

Hagelberg, E. (1997) 'Ancient and modern mitochondrial DNA sequences and the colonization of the Pacific', *Electrophoresis* 18 (9): 1529–33.

Hagelberg, E. and Clegg, J. B. (1993) 'Genetic polymorphisms in prehistoric Pacific Islanders determined by analysis of ancient bone DNA', *Proceedings of the Royal Society of London*, Series B, 252: 163–70.

Hagelberg, E. *et al.* (1994) 'DNA from ancient Easter Islanders', *Nature* 369: 25–6.

Hagelberg, E. *et al.* (1999) 'Evidence for mitochondrial DNA recombination in a human population of island Melanesia', *Proceedings of the Royal Society of London*, Series B, 266: 485–92.

Hagelberg, E. *et al.* (1999) 'Molecular genetic evidence for human settlement of the Pacific: analysis of mitochondrial DNA, Y chromosome and HLA markers', *Philosophical Transactions of the Royal Society of London*, Series B, 354: 141–52.

Harpending, H. C. *et al.* (1993) 'The genetic structure of ancient human populations', *Current Anthropology* 34 (4): 483–96.

Harpending, H. C. *et al.* (1998) 'Genetic traces of ancient demography', *Proceedings of the National Academy of Science USA* 95: 1961–7.

Hauswirth, W. W., Dickel, C. D. and Lawlor, D. A. (1994) 'DNA analysis of the Windover population', in B. Herrmann and S. Hummel (eds), *Ancient DNA. Recovery and Analysis of Genetic Material from Palaeontological, Archaeological, Museum, Medical and Forensic Specimens*, New York: Springer-Verlag.

Heyerdahl, T. and Ferdon, E. N. (eds) (1966) *Report of the Norwegian Archaeological Expedition to Easter Island and the East Pacific*, London: Allen and Unwin.

Hill, A. V. S. and Serjeantson, S. W. (eds) (1989) *The Colonisation of the Pacific: A Genetic Trail*. Oxford: Clarendon Press.

Horai, S. *et al.* (1993) 'Peopling of the Americas, founded by four major lineages of mitochondrial DNA', *Molecular Biology and Evolution* 10: 23–47.

Hurles, M. E. *et al.* (1999) 'European mitochondrial lineages in Polynesia: a contrast to the population structure revealed by mitochondrial DNA', *American Journal of Human Genetics* 63: 1793–1806.

Krings, M. *et al.* (1999) 'mtDNA analysis of Nile River Valley populations: a genetic corridor or a barrier to migration?', *American Journal of Human Genetics* 64: 1166–76.

Lalueza Fox, C. (1996) 'Analysis of ancient mitochondrial DNA from extinct Aborigines from Tierra del Fuego–Patagonia', *Ancient Biomolecules* 1 (1): 43–54.

— (1997) 'mtDNA analysis in ancient Nubians supports the existence of gene flow between sub-Sahara and North Africa in the Nile Valley', *Annals of Human Biology* 24 (3): 217–27.

Lum, J. K. and Cann, R. L. (1998) 'mtDNA and language support a common origin of Micronesians and Polynesians in Island Southeast Asia', *American Journal of Physical Anthropology* 105 (2): 109–19.

Lum, J. K. *et al.* (1998) 'Mitochondrial and nuclear genetic relationships among Pacific Island and Asian populations', *American Journal of Human Genetics* 63: 613–24.

Merriwether, A. D. and Ferrell, R. E. (1996) 'The four founding lineage hypothesis for the New World: a critical re-evaluation', *Molecular Phylogeny and Evolution* 5: 241–6.

Merriwether, A. D., Rothhammer, F. and Ferrell, R. (1994) 'Genetic variation in the new world: ancient teeth, bone, and tissue as source of DNA', *Experientia* 50 (6): 592–601.

Merriwether, A. D. *et al.* (1996) 'mtDNA variation indicates Mongolia may have been the source for the founding population for the New World', *American Journal of Human Genetics* 59: 204–12.

Monsalve, M. V. (1997) 'Mitochondrial DNA in ancient Amerindians', *American Journal of Physical Anthropology* 103 (3): 423–5.

Parr, R. L., Carlyle, S. W. and O'Rourke, D. H. (1996) 'Ancient DNA analysis of Fremont Amerindians of the Great Salt Lake Wetlands', *American Journal of Physical Anthropology* 99 (4): 507–18.

Renfrew, A. C. (1987) *Archaeology and Language: The Puzzle of Indo-European Origins*, London: Jonathan Cape.

— (1994) 'World linguistic diversity', *Scientific American* 207 (1): 116–23.

— (ed.) (2000) *America Past, America Present: Genes and Languages in the Americas and Beyond*, Cambridge: McDonald Institute Monograph.

Ribeiro-dos-Santos, A. K. *et al.* (1996) 'Heterogeneity of mitochondrial DNA haplotypes in pre-Columbian natives of the Amazon region', *American Journal of Physical Anthropology* 101 (1): 29–37.

Richards, M. *et al.* (1996) 'Paleolithic and Neolithic lineages in the European mitochondrial gene pool', *American Journal of Human Genetics* 59: 185–203.

Sajantila, A. *et al.* (1995) 'Genes and languages in Europe: an analysis of mitochondrial lineages', *Genome Research* 5 (1): 42–52.

Semino, O. *et al.* (1989) 'Mitochondrial DNA polymorphisms in Italy', III, 'Population data from Sicily: a possible quantitation of maternal African ancestry', *Annals of Human Genetics* 53: 193–202.

Smith, G. E. (1915) *The Migrations of Early Culture: A Study of the Significance of the Geographical Distribution of the Practice of Mummification as Evidence of the Migration of Peoples and the Spread of Certain Customs and Beliefs*, Manchester: Manchester University Press.

— (1933) *The Diffusion of Culture*, London: Watts and Co.

Spriggs, M. (1984) 'The Lapita cultural complex', in R. L. and E. Szathmary (eds), *Out of Asia*, Canberra: Australian National University Press, pp. 185–202.

Starikovskaya, Y. B. *et al.* (1998) 'mtDNA diversity in Chukchi and Siberian Eskimos: implications for the New World', *American Journal of Human Genetics* 63: 1473–91.

Stone, A. C. and Stoneking, M. (1993) 'Ancient DNA from a pre-Columbian Amerindian population', *American Journal of Physical Anthropology* 92: 463–71.

— and — (1996) 'Genetic analyses of an 8,000-year-old native American skeleton', *Ancient Biomolecules* 1 (1): 83–7.

— and — (1998) 'mtDNA analysis of a prehistoric Oneota population: implications for the peopling of the New World', *American Journal of Human Genetics* 62: 1153–70.

— and — (1999) 'Analysis of ancient DNA from a prehistoric Amerindian cemetery', *Philosophical Transactions of the Royal Society of London*, Series B, 354: 153–9.

Sykes, B. C. (1999) 'The molecular genetics of European ancestry', *Philosophical Transactions of the Royal Society*, Series B, 354: 131–9.

Sykes, B. C. *et al.* (1995) 'The origins of the Polynesians: an interpretation from mitochondrial lineage analysis', *American Journal of Human Genetics* 57: 1463–75.

Torroni, A. *et al.* (1993) 'Asian affinities and continental radiation of the four founding Native American mtDNAs', *American Journal of Human Genetics* 53: 563–90.

Wallace, D. C. and Torroni, A. (1992) 'American Indian prehistory as written in the mitochondrial DNA: a review', *Human Biology* 64 (3): 403–16.

Wallace, D. C., Garrison, G. and Knowler, W. C. (1985) 'Dramatic founder effects in Amerindian mitochondrial DNAs', *American Journal of Physical Anthropology* 68: 149–55.

Ward, R. H. *et al.* (1991) 'Extensive mitochondrial diversity within a single Amerindian tribe', *Proceedings of the National Academy of Science USA* 88: 8720–4.

West, F. H. and West, C. F. (1996) *American Beginnings: The Prehistory and Palaeoecology of Beringia*, Chicago: University of Chicago Press.

8 beyond DNA

Allison, M. J., Castro, N. and Hosseini, A. (1976) 'ABO blood groups in Peruvian mummies', *American Journal of Physical Anthropology* 44: 55–62.

Ascenzi, A. *et al.* (1985) 'Immunological detection of hemoglobin in bones of ancient Roman times and Iron and Neolithic ages', *Proceedings of the National Academy of Science USA* 82: 7170–2.

Balasse, M., Bocherens, H. and Mariotti, A. (1999) 'Intra-bone variability of collagen and apatite isotopic composition used as evidence of a change of diet', *Journal of Archaeological Science* 26 (6): 593–8.

Bocherens, H. *et al.* (1991) 'Isotopic biogeochemistry (13C, 15N) of fossil vertebrate collagen: application to the study of a past food web including Neandertal man', *Journal of Human Evolution* 20: 481–92.

Borgognini, S. M. *et al.* (1979) 'On the possibility of the MN blood group determination in human bones', *Journal of Human Evolution* 8: 725–34.

Boyd, W. C. (1959) 'A possible example of the action of selection in human blood groups?' *Journal of Medical Education* 34: 398–9.

Boyd, W. C. and Boyd, L. G. (1933) 'Blood grouping by means of preserved muscle', *Science* 78: 578.

— and — (1937) 'Blood group testing on 800 mummies', *Journal of Immunology* 32: 307–16.

Briggs, D. E. (1999) 'Molecular taphonomy of animal and plant cuticles: selective preservation and diagenesis', *Philosophical Transactions of the Royal Society of London*, Series B, 354: 7–17.

Browne, T. (1658) *Hydriotaphia, Urne-Buriall, or, A Discourse of the Sepulchrall Urnes lately found in Norfolk*, London: 'Printed for Hen. Browne at the Sign of the Gun in Ivy-Lane'.

Cattaneo, C. *et al.* (1990) 'Blood in ancient human bone', *Nature* 347: 339.

Cattaneo, C. *et al.* (1991) 'Identification of ancient blood and tissue – ELISA and DNA analysis', *Antiquity* 65: 878–81.

Cattaneo, C. *et al.* (1992) 'Detection of human proteins in buried blood using ELISA and monoclonal antibodies: towards the reliable species identification of blood stains on buried material', *Forensic Science International* 57 (2): 139–46.

Cattaneo, C. *et al.* (1992) 'Reliable identification of human albumin in ancient bone using ELISA and monoclonal antibodies', *American Journal of Physical Anthropology* 87 (3): 365–72.

Cattaneo, C. *et al.* (1993) 'Blood residues on stone tools: indoor and outdoor experiments', *World Archaeology* 25 (1): 29–43.

Cattaneo, C. *et al.* (1995) 'Differential survival of albumin in ancient bone', *Journal of Archaeological Science* 22 (2): 271–6.

Chrisholm, B. S., Nelson, D. E. and Schwarz, H. P. (1983) 'Dietary information from delta C13 and delta N15 measurements', *PACT* 8: 391–5.

Collins, M. J. *et al.* (1999) 'Is osteocalcin stabilised in ancient bones by adsorption to bioapatite?' *Ancient Biomolecules* 2 (2): 223–33.

Collinson, M. E. *et al.* (1999) 'The preservation of plant cuticle in the fossil record: a chemical and microscopical investigation', *Ancient Biomolecules* 2 (2): 251–65.

Curry, G. B. (1987) 'Molecular palaeontology: new life for old molecules', *Trends in Ecology and Evolution* 2 (6): 161–5.

Curry, G. B. *et al.* (1991) 'Biochemistry of brachiopod intracrystalline molecules', *Philosophical Transactions of the Royal Society of London*, Series B, 333: 359–66.

Dickson, J. H. *et al.* (2000) 'The omnivorous Tyrolean Iceman: colon contents (meat, cereals, pollen, moss and whipworm) and stable isotope analyses', *Philosophical Transactions of the Royal Society of London*, Series B, 355: 1843–9.

Eglinton, G. and Curry, G. B. (eds) (1991) *Molecules through Time: Fossil Molecules and Biochemical Systematics*, London: Royal Society.

Evershed, R. P. (1990) 'Preliminary report of the analysis of lipids from samples of skin from seven Dutch bog bodies', *Archaeometry* 33: 139–53.

— (1993) 'Biomolecular archaeology and lipids', *World Archaeology* 25 (1): 74–93.

Evershed, R. P. *et al.* (1995) 'Preliminary results for the analysis of lipids in ancient bone', *Journal of Archaeological Science* 22: 277–90.

Evershed, R. P. *et al.* (1999) 'Lipids as carriers of anthropogenic signals from prehistory', *Philosophical Transactions of the Royal Society of London*, Series B, 354: 19–31.

Gurfinkel, D. M. and Franklin, U. M. (1988) 'A study of the feasibility of detecting blood residue on artifacts', *Journal of Archaeological Science* 15: 83–97.

Hastorf, C. A. and DeNiro, M. J. (1985) 'Reconstruction of prehistoric plant production and cooking practices by a new isotopic method', *Nature* 315: 489–91.

Hirschfeld, L. and Hirschfeld, H. (1919) 'Serological differences between the blood of different races', *Lancet* 2: 675–8.

Hyland, D. C. and Tersak, J. M. (1990) 'Identification of the species of origin of residual blood on lithic material', *American Antiquity* 55: 104–12.

Jahren, A. H. *et al.* (1997) 'Determining stone tool use: chemical and morphological analyses of residues on experimentally manufactured stone tools', *Journal of Archaeological Science* 24 (3): 245–50.

Lowenstein, J. M. (1980) 'Species-specific proteins in fossils', *Naturwissenschaften* 67 (7): 343–6.

— (1981) 'Immunological reactions from fossil material', *Philosophical Transactions of the Royal Society of London*, Series B, 292: 143–9.

— (1985) 'Radioimmune assay of mammoth tissue', *Acta Zoologica Fennica* 170: 233–5.

Lowenstein, J. M. and Ryder, O. A. (1985) 'Immunological systematics of the extinct quagga (Equidae)', *Experientia* 41 (9): 1192–3.

Lowenstein, J. M., Sarich, V. M. and Richardson, B. J. (1981) 'Albumin systematics of the extinct mammoth and Tasmanian wolf', *Nature* 291: 409–11.

Lowenstein, J. M. and Scheuenstuhl, G. (1991) 'Immunological methods in molecular palaeontology', *Philosophical Transactions of the Royal Society of London*, Series B, 333: 375–80.

Loy, T. H. (1983) 'Prehistoric blood residues: detection on tool surfaces and identification of species of origin', *Science* 220: 1269–70.

— (1991) 'Prehistoric organic residues: recent advances in identification, dating, and their antiquity', in E. Pernicka and G. A. Wagner (eds), *Archaeometry '90*, Basel: Birkhauser Verlag, pp. 645–56.

— (1992) 'Detection, amplification and identification of 2,800-year-old DNA from blood residues on prehistoric stone tools', *Ancient DNA Newsletter* 1 (2): 20.

— (1998) 'Blood on the axe', *New Scientist* (12 Sept.): 40–3.

Loy, T. H. and Matthaei, K. I. (1994) 'Species of origin determination from prehistoric blood residues using ancient genomic DNA', *Australasian Biotechnology* 4 (3): 161–2.

Loy, T. H., Spriggs, M. and Wickler, S. (1992) 'Direct evidence for human use of plants 28,000 years ago: starch residues on stone artifacts from the northern Solomons', *Antiquity* 66: 898–912.

Macko, S. A. *et al.* (1999) 'Documenting the diet in ancient human populations through stable isotope analysis of hair', *Philosophical Transactions of the Royal Society of London*, Series B, 354: 65–76.

Newman, M. E. *et al.* (1997) '"Blood" from stones? Probably: a response to Fiedel', *Journal of Archaeological Science* 24 (11): 1023–7.

Piperno, D. R. (1988) *Phytolith Analysis: An Archaeological and Geological Perspective*, New York: Academic Press.

Piperno, D. R. and Holst, I. (1998) 'The presence of starch grains on prehistoric stone tools from the humid neotropics: indications of early tuber use and agriculture in Panama', *Journal of Archaeological Science* 25 (8): 765–76.

Smith, R. and Wilson, M. T. (1990) 'Detection of haemoglobin in human skeletal remains by ELISA', *Journal of Archaeological Science* 17: 255–68.

Springfield, A. C. *et al.* (1993) 'Cocaine and metabolites in the hair of ancient Peruvian coca leaf chewers', *Forensic Science International* 63: 269–75.

Stankiewicz, A. B. *et al.* (1997) 'Preservation of chitin in 25-million-year-old fossils', *Science* 276: 1541–3.

Stott, A. W. *et al.* (1999) 'Cholesterol as a new source of palaeodietary information: experimental approaches and archaeological applications', *Journal of Archaeological Science* 26 (6): 705–16.

Sullivan, C. H. and Krueger, H. W. (1983) 'Carbon isotope ratios of bone apatite and animal diet reconstruction', *Nature* 301: 177–8.

Tuross, N. and Stathoplos, L. (1993) 'Ancient proteins in fossil bones', *Methods in Enzymology* 224: 121–9.

Van der Merwe, N. J. (1992) 'Light stable isotopes and the reconstruction of prehistoric diets', in M. Polland (ed.), *New Developments in Archaeological Science*, Oxford: Oxford University Press, pp. 247–64.

9 friends and relations

Colson, I. B. *et al.* (1997) 'DNA analysis of seven human skeletons excavated from the Terp of Wijnaldum', *Journal of Archaeological Science* 24 (10): 911–17.

Corach, D. *et al.* (1997) 'Additional approaches to DNA typing of skeletal remains: the search for "missing" persons killed during the last dictatorship in Argentina', *Electrophoresis* 18: 1608–12.

Delefosse, T. and Hänni, C. (1997) 'Archéologie moléculaire: relation de parenté au sein d'un gisement néolithique', *Comptes rendus des séances de la Société de Biologie et de ses Filiales* 191 (4): 521–8.

Gerstenberger, J., Hummel, S. and Herrmann, B. (1998) 'Assignment of an isolated skeletal element to the skeleton of Duke Christian II through STR-typing', *Ancient Biomolecules* 2 (1): 63–8.

Gerstenberger, J. *et al.* (1999) 'Reconstruction of a historical genealogy by means of STR analysis and Y-haplotyping of ancient DNA', *European Journal of Human Genetics* 7 (4): 469–77.

Gill, P. *et al.* (1994) 'Identification of the remains of the Romanov family by DNA analysis', *Nature Genetics* 6: 130–5.

Hagelberg, E., Gray, I. C. and Jeffreys, A. J. (1991) 'Identification of the skeletal remains of a murder victim by DNA analysis', *Nature* 352: 427–9.

Hauswirth, W. W. *et al.* (1994) 'Inter- and intrapopulation studies of ancient humans', *Experientia* 50 (6): 585–91.

Herrmann, B. (1997) 'Prospects in prehistoric anthropology?' *Anthropologischer Anzeiger* 55 (2): 97–100.

Hummel, S., and Herrmann, B. (1991) 'Y-chromosome-specific DNA amplified in ancient human bones', *Naturwissenschaften* 78: 266–7.

— and — (1996) 'aDNA typing for reconstruction of kinship', *Homo* 47 (1–3): 215–22.

— and — (1997) 'Verwandtschaftsfeststellung durch aDNA-analyse', *Anthropologischer Anzeiger* 55 (2): 217–23.

Ivanov, P. *et al.* (1996) 'Mitochondrial DNA sequence heteroplasmy in the Grand Duke of Russia Georgij Romanov establishes the authenticity of the remains of Tsar Nicholas II', *Nature Genetics* 12: 417–20.

Jeffreys, A. J., Wilson, V. and Thein, S. L. (1985) 'Individual-specific "fingerprints" of human DNA', *Nature* 316: 76–9.

Jeffreys, A. J. *et al.* (1992) 'Identification of the skeletal remains of Josef Mengele by DNA analysis', *Forensic Science International* 56 (1): 65–76.

Lassen, C. S., Hummel, S. and Herrmann, B. (1996) 'PCR-based sex identification of ancient human bones by amplification of X- and Y-chromosomal sequences: a comparison', *Ancient Biomolecules* 1 (1): 25–33.

Lawlor, D. A. *et al.* (1991) 'Ancient HLA genes from 7,500-year-old archaeological remains', *Nature* 349: 785–7.

Lingxia, Z. *et al.* (1996) 'Ancient DNA extraction from Neolithic human remains and PCR-based amplification of the X-Y-homologous amelogenin gene', *Acta Anthropologica Sinica* 15: 206–9.

Macko, S. A. *et al.* (1999) 'Documenting the diet in ancient human populations through stable isotope analysis of hair', *Philosophical Transactions of the Royal Society of London*, Series B, 354: 65–76.

Murdock, G. P. (1949) *Social Structure*, New York: Macmillan.

Oota, H. *et al.* (1995) 'A genetic study of 2,000-year-old human remains from Japan using mitochondrial DNA sequences', *American Journal of Physical Anthropology* 98 (2): 133–45.

Oota, H. *et al.* (1999) 'Molecular genetic analysis of remains of a 2,000-year-old human population in China and its relevance for the origin of the modern Japanese population', *American Journal of Human Genetics* 64: 250–8.

Primorac, D. *et al.* (1996) 'Identification of war victims from mass graves in Croatia, Bosnia, and Herzegovina by use of standard forensic methods and DNA typing', *Journal of Forensic Science* 41 (5): 891–4.

Stone, A. C. and Stoneking, M. (1998) 'mtDNA analysis of a prehistoric Oneota population: implications for the peopling of the New World', *American Journal of Human Genetics* 62: 1153–70.

— and — (1999) 'Analysis of ancient DNA from a prehistoric Amerindian cemetery', *Philosophical Transactions of the Royal Society of London*, Series B, 354: 153–9.

Thurnwald, R. C. (1932) *Die Menschliche Gesellschaft in ihren ethnisoziologischen Grundlagen II*, Berlin, pp. 193–4.

10 enemies within

Arriaza, B. T. *et al.* (1995) 'Pre-Columbian tuberculosis in northern Chile: molecular and skeletal evidence', *American Journal of Physical Anthropology* 98 (1): 37–45.

Baron, H., Hummel, S. and Herrmann, B. (1996) '*Mycobacterium Tuberculosis* complex DNA in ancient human bones', *Journal of Archaeological Science* 23: 667–71.

Cano, R. J. and Borucki, M. K. (1995) 'Revival and identification of bacterial spores in 25- to 40-million-year-old Dominican amber', *Science* 268: 1060–64.

Dawes, J. D. and Magilton, J. R. (1980) *The Cemetery of St Helen-on-the-Walls, Aldwark*, London: Council for British Archaeology.

Donoghue, H. *et al.* (1998) *Mycobacterium tuberculosis* complex DNA in calcified pleura from remains 1,400 years old', *Letters in Applied Microbiology* 27 (5): 265–9.

Dormandy, T. (1999) *The White Death: A History of Tuberculosis*, London: Hambledon Press.

Drancourt, M. *et al.* (1998) 'Detection of 400-year-old *Yersinia pestis* DNA in human dental pulp: an approach to the diagnosis of ancient septicemia', *Proceedings of the National Academy of Science USA* 95: 12637–40.

Faerman, M. *et al.* (1997) 'Prevalence of human tuberculosis in a medieval population of Lithuania studied by ancient DNA analysis', *Ancient Biomolecules* 1 (3): 205–14.

Fricker, E. J., Spigelman, M. and Fricker, C. R. (1997) 'The detection of *Escherichia coli* DNA in the ancient remains of Lindow man using the polymerase chain reaction', *Letters in Applied Microbiology* 24 (5): 351–4.

Gernaey, A. M. *et al.* (1998) 'Detecting ancient tuberculosis', *Internet Archaeology* 5.

Guhl, F. *et al.* (1999) 'Isolation of *Trypanosoma cruzi* DNA in 4,000-year-old mummified human tissue from northern Chile', *American Journal of Physical Anthropology* 108 (4): 401–7.

Handt, O. *et al.* (1994) 'Molecular genetic analyses of the Tyrolean ice man', *Science* 264: 1775–8.

Nerlich, A. G. *et al.* (1997) 'Molecular evidence for tuberculosis in an ancient Egyptian mummy', *The Lancet* 350 (9088): 1404.

Priest, F. G., Beckenbach, A. T. and Cano, R. J. (1995) 'Age of bacteria from amber', *Science* 270: 2015–17.

Rafi, A. *et al.* (1994) '*Mycobacterium leprae* DNA from ancient bone detected by PCR', *The Lancet* 343 (8909): 1360–1.

Rhodes, A. N. *et al.* (1998) 'Identification of bacterial isolates obtained from intestinal contents associated with 12,000-year-old mastodon remains', *Applied Environmental Microbiology* 64 (2): 651–8.

Roberts, C. and Manchester, K. (1995) *The Archaeology of Disease*. 2nd edn, Stroud: Sutton.

Rollo, F. and Marota, I. (1999) 'How microbial ancient DNA, found in association with human remains, can be interpreted', *Philosophical Transactions of the Royal Society of London*, Series B, 354: 111–19.

Rollo, F., Sassaroli, S. and Ubaldi, M. (1995) 'Molecular phylogeny of the fungi of the Iceman's grass clothing', *Current Genetics* 28: 289–97.

Rollo, F. *et al.* (1994) 'Molecular ecology of a Neolithic meadow', *Experientia* 50: 576–84.

Rollo, F. *et al.* (1995) 'The "neolithic" microbial flora of the Iceman's grass: morphological description and DNA analysis', in K. Spindler *et al.* (eds), *Der Mann im Eis*, Vol. 2, Springer-Verlag: Vienna and New York, pp. 107–14.

Salo, W. L. *et al.* (1994) 'Identification of *Mycobacterium tuberculosis* DNA in a pre-Columbian Peruvian mummy', *Proceedings of the National Academy of Science USA* 91: 2091–4.

Spigelman, M. and Lemma, E. (1993) 'The use of the polymerase chain reaction (PCR) to detect *Mycobacterium tuberculosis* in ancient skeletons, *International Journal of Osteoarchaeology* 3: 137–43.

Taylor, G. M. (1996) 'DNA from *Mycobacterium tuberculosis* identified in mediaeval human skeletal remains using polymerase chain reaction', *Journal of Archaeological Science* 23: 789–98.

Taylor, G. M., Rutland, P. and Molleson, T. (1997) 'A sensitive polymerase chain reaction method for the detection of *Plasmodium* species DNA in ancient human remains', *Ancient Biomolecules* 1 (3): 193–203.

Taylor, G. M. *et al.* (1999) 'Genotypic analysis of *Mycobacterium tuberculosis* from medieval human remains', *Microbiology* 145 (4): 899–904.

Thierry, D. *et al.* (1990) 'IS6110, an IS-like element of *Mycobacterium tuberculosis* complex', *Nucleic Acids Research* 181: 188.

Ubaldi, M., Sassaroli, S. and Rollo, F. (1996) 'Ribosomal DNA analysis of culturable deuteromyces from the Iceman's hay: comparison of living and mummified fungi', *Ancient Biomolecules* 1 (1): 35–42.

Ubaldi, M. *et al.* (1998) 'Sequence analysis of bacterial DNA in the colon of an Andean mummy', *American Journal of Physical Anthropology* 107 (3): 285–95.

11 the hunt goes on

Clarke, D. L. (1972) 'A provisional model of an Iron Age society and its settlement system', in D. L. Clarke (ed.), *Models in Archaeology*, London: Methuen, pp. 801–69.

Collins, M. J., Waite, E. R. and Van Duin, A. C. T. (1999) 'Predicting protein decomposition: the case of aspartic-acid racemization kinetics', *Philosophical Transactions of the Royal Society of London*, Series B, 354: 51–64.

Geigl, E.-M. (1996) '*Homo erectus* of Western France was not a vegetarian: zoological assignment of a fossil bone of Menez-Dregan I via a palaeogenetical approach', in *Actes du XIIIe congres UISPP*, section 5.3.

— (1997) 'L'emergence de la paleogénétique', *Biofutur* 164: 28–34.

Underhill, A. *et al.* (1996) 'A pre-Columbian Y chromosome-specific transition and its implications for human evolutionary history', *Proceedings of the National Academy of Science USA* 93: 196–200.

index